PLAY WITH YOUR CAT!

The Essential Guide to Interactive Play for a Happier, Healthier Feline

MIKEL MARIA DELGADO, PhD

ILLUSTRATIONS BY LILI CHIN

A TARCHERPERIGEE BOOK

An imprint of Penguin Random House LLC
penguinrandomhouse.com

Most TarcherPerigee books are available at special quantity discounts for bulk purchase for sales promotions, premiums, fundraising and educational needs. Special books or book excerpts also can be created to fit specific needs. For details, write SpecialMarkets@penguinrandomhouse.com.

Library of Congress Cataloging-in-Publication Data

Names: Delgado, Mikel, author.
Title: Play with your cat!: the essential guide to interactive play for a happier, healthier feline / Mikel Maria Delgado, PhD; illustrations by Lili Chin.
Description: [New York]: TarcherPerigee,
Penguin Random House LLC, [2024] | Includes bibliographical references.
Identifiers: LCCN 2023030786 (print) | LCCN 2023030787 (ebook) |
ISBN 9780593541333 (trade paperback) | ISBN 9780593541340 (epub)
Subjects: LCSH: Cats—Exercise. | Games for cats. |
Cats—Behavior. | Cats—Training.
Classification: LCC SF446.7.D45 2024 (print) | LCC SF446.7 (ebook) |
DDC 636.8/083—dc23/eng/20230804
LC record available at https://lccn.loc.gov/2023030786
LC ebook record available at https://lccn.loc.gov/2023030787

Printed in the United States of America
1st Printing

DEDICATED TO SCOTT, FOR ALWAYS
SUPPORTING ME AS I'VE FOLLOWED
MY FELINE DREAMS, AND TO CLARABELLE

CONTENTS

PLAY WITH YOUR CAT!

CHAPTER 1

Why Play?

A rustling sound awakens the cat from his nap on the couch. He raises his head, a little bleary from sleep. His eyes slowly focus on the bird in the next room gently hopping among the leaves (or is that tissue paper?) on the floor. Suddenly alert, he rolls into an upright position and creeps to a place just behind the arm of the couch, hiding from the bird's view as he continues to observe. The bird begins to hop and flit about, and the cat silently slinks off the couch and quickly takes shelter behind a large potted houseplant, which he perceives as a shrub. He drops low to the ground, which strangely is carpeted. His pupils dilate and his ears point forward, taking in as much sensory information as he can to ensure he is prepared to catch his prey. As he plans his final attack, his back legs begin to tread and his tail twitches forcefully. He makes a leap, springing then landing just in front of the bird, simultaneously striking the feathery creature with his front paws.

The bird struggles, and the cat flops onto his side, gripping with his front paws and biting while kicking the feathers with his back feet. The cat glances up and notices his human standing nearby, holding a stick with a string attached. The string is connected to . . . the bird? How can that be? In this moment of distraction, the bird escapes his grip and flies away. The cat chases the bird, leaping into the air in hot pursuit, with all four paws off the ground, arcing into a perfect backflip, and expertly recapturing the bird.

The cat considers how odd everything seems. He's not outdoors after all, but in the living room of his home. Why is there a bird in the living room? And why is the bird attached to his human? On the other hand, the toy looks like a bird, it feels like a bird, and it's acting like a bird. Is there any harm in pretending it's a bird? The cat may realize this is play, but he also knows this feels almost as good as hunting.

♥ ♥ ♥

Play follows humans throughout our lives. When we are children, we craft toys out of rocks and sticks, but we are also engrossed by the bright and shiny toys that come in packages. We pretend to build, cook, or perform surgery. We care for our stuffed animals as if they were children, nursing them to health or putting them to bed. We learn games and play or watch group sports. Play is considered important, even essential, to children's development, creativity, relationships, and well-being.

As adults, many of us still enjoy board games, video games, or sports. We tell stories, make jokes, and create art. We interact with the world around us, using our imaginations, testing out relationships or skills, and most of the time we're also having fun along the way. But as we age, we often deprioritize play because we have so many responsibilities we must attend to. Those bills don't pay themselves.

In cases where children aren't given the opportunity to play freely (without adult control), anxiety and depression increase. And for adults, the adage "all work and no play makes Jack a dull boy" exists for a reason: without play, we become bored and boring.

FOREVER YOUNG

When play isn't high on the priority list for us, we tend to minimize its importance to our companion animals. But our cats and dogs (and other pets) don't "grow up"—they are never going to take care of themselves and leave home and get a job! Through the process of domestication, we have selected cats for neoteny, or infant-like features, such as large eyes, a round face, a shorter nose, and likely even for friendly behaviors like cuddling and meowing.

When we spay and neuter our pets, we also block adult sex hormones from circulating throughout the bloodstream. By doing so, we've increased the odds they will retain juvenile behaviors, making them less likely to express some "adult" tendencies (such as roaming and marking their territory). We call them "fur babies" for a reason. As a result, our pets remain perpetually young at heart to an extent, and that includes a lifelong desire to play!

PLAY TO LIVE

When we think of behaviors that are important for survival (especially in other species), we tend to focus on mating, acquiring food, defending territory, and avoiding being eaten by other animals. Play doesn't often rank high on that list.

Perhaps this is because the word *play* implies frivolity: wasted time, something that only children do, something that we do when the serious parts of life (work, school, hygiene, sleep) are taken care of. But what about when we consider play as creative and imaginative, as practice for the "real world," to safely test situations that might otherwise be risky?

Play can help animals grow into competent adults. Animals can improve their motor skills by interacting with structures and three-dimensional spaces while exploring their environments (locomotor play). Animals can learn how to acquire food or hunt by interacting with objects (object play). Young animals often learn how to interact with other individuals through rough-and-tumble activities (social play).

When we look at play through this functional lens, animal play begins to make sense. That said, play might not *always* serve a greater purpose—I have no doubt that sometimes play's function is nothing more than "it is fun" or "it feels good."

Some research has even shown that animals play when they are stressed out—making play a bit of an outlet for that stress, or a way to cope. For some it's preventive medicine, and for others it's a cure. Neuroscientists have pointed to the fact that when animals play, parts of the brain related to motivation, emotions, and reward are activated. Some animals play when they feel good, and others might play *to* feel good.

PLAYING MAKES YOU SMART

Just as play may provide cognitive benefits for humans, play in other animals is associated with larger brains—the more elaborate a species' play life is, the larger the relative brain size they have. Although "bird smarts" are often attributed to their ability to use tools, it turns out that it is not tool use that predicts bird brain size but play. Birds who engaged in social play had the largest brains of all, but even the birds who engaged only in object play had measurably larger brains than those of birds who did not play at all.

Why would play help brains? Play and exercise are associated with increases in the amount of gray matter, tissue that is predominant in areas of the brain (such as the cerebral cortex and cerebellum) that are associated with learning, memory, and coordination. Play may protect brain cells from dying or, in some cases, may even promote growth or reassignment of existing brain cells. Those same feel-good brain chemicals that are released during play also aid in learning, further cementing life lessons and making the brain more receptive to absorbing new information.

WHY PLAY

BENEFITS FOR ALL

- Prepares individuals for "real-life" experiences
- Promotes brain growth and health
- Feels good
- Improves mental health
- Increases creativity

PLAY: IT'S GOOD FOR ALL OF US

There are good reasons that play is found throughout the animal kingdom. It can help with depression and anxiety, it can help prepare us for life experiences, and it may even make us smarter. All of these benefits are likely true for our cats as well. Just as importantly, as we'll learn in chapter 6, an absence of play can be a warning sign that all is not well with your cat or their environment.

Although the focus of this book is how to improve your cat's life through play, my motives are many here. I hope this book, through encouraging a fresh look at how you play with your cat, will help you have a better relationship with them. I know that when I'm playing with my cats there are lots of smiles and laughs on my end, so I'd like to think this book could improve your life as well. Play lightens our hearts and can help take our minds off things that are bothering us. Play is fun for you and it's fun for your cat. Play is good for you and it's good for your cat.

Let's begin, shall we?

CHAPTER 2

Cats Play Because They Hunt!

We don't like to think about the fact that our cute, purring, cuddly cats are actually stone-cold killers. But it is a fact: Cats hunt. Without hunting, they would never have survived as a species, for it was how they fed themselves. It is a hardwired instinct in the core being of all cats. No, we don't have to let them hunt. Yes, it is good to keep them far from our bird feeders. But we cannot take their desire to hunt away from them. It doesn't go away just because we may have clipped their nails and kept them in our apartments and homes. Hunting makes them happy. And for cats, hunting is the root of playing.

To understand how to play with our cats, we need to understand how they hunt. There are a lot of misconceptions and mysteries surrounding cat-hunting behavior (Do they torture their prey? Do they bring me dead animals because they think I am hungry?); and to be certain, there are individual differences between cats in how skilled and interested they are in hunting.

Research shows that hunting is an innate behavior that almost all cats will put into practice given the opportunity, regardless of lifestyle or previous experience. Even cats who are cared for and fed by humans will hunt and kill given the chance, despite not always eating what they catch. And although it may not be readily apparent when your cat is lounging belly-up on the couch, the killer instinct is just below the surface, waiting to reveal itself.

Luckily there are plenty of ways to engage with that instinct—without bloodshed—through interactive play!

HOW AN ASSASSIN IS BORN

Kittens are born *altricial*, meaning they are helpless and depend on Mom for warmth, safety, and food. (Precocial animals are born ready to go, without a need for maternal care. Guinea pigs and horses are two such examples.) Although kittens are born blind, deaf, and only able to wiggle around, within a matter of months, they grow into competent, self-sufficient hunters.

How do kittens get from A to B? During their first few weeks, kittens spend most of their time nursing and sleeping. They are guided primarily by the senses of smell and touch. By two weeks of age, kittens' eyes and ears are functioning, allowing them to begin to interact with the world around them. In the third week of life, kittens are becoming more coordinated, helping them to engage in play with littermates and small objects. Many of these frisky interactions resemble aspects of hunting behavior. When her kittens are around a month old, the mother will start the process of weaning. Multiple changes will occur for kittens during this time:

- Mom introduces her kittens to the concept of meat by eating prey in front of them.
- Next, she will present her kittens with dead prey.
- Kittens begin the dietary transition from mother's milk to meat.
- Mom spends less time with her kittens, hastening their independence and development.

Mom will up the game over the next few weeks by bringing home weakened prey for her kittens, to let them practice killing. Although kittens instinctively pursue and chase prey, their hunting understandably lacks finesse at first. Mom may have to offer the occasional assist while kittens learn the fine art of mouse dismemberment.

Folklore says that kittens learn to hunt from Mom, and if they don't, they will not succeed as predators. The evidence suggests this is not true. Watching Mom hunt gives kittens a leg up, but it is not sufficient or even necessary for a kitten to grow into a successful killer. Even kittens who are prematurely separated from their mothers can still hunt successfully through instinct and with practice. Though time under Mom's tutelage is a benefit, hunting is so important to feline survival that practice will eventually make perfect (or at least halfway decent) even without a good instructor in charge!

With adequate practice on live prey, an adept hunter can deliver one sharp bite to the nape of their prey's neck, leading to a swift, orderly kill. When a cat meets their mark, their canine teeth slide between the vertebrae of the neck like a lock and key, severing the spinal cord of their victim. It is not just about saving time; failure to kill quickly allows a mouse or rat to escape or even fight back, increasing the chance that a cat might get injured or go hungry.

Kittens can't predict the future, and they don't know that they may spend their days as a house cat with humans to care for them. So, for the sake of survival, kittens must practice, and practice, the fine art of delivering that killing bite.

♥ ♥ ♥

Some predators will chase their prey for extensive periods, in what is known as *pursuit or persistence predation*. Persistence predators—such as cheetahs or wolves—run long distances, waiting for their prey to tire. They take advantage of that fatigue to make their final attack. Cats go for a much more laid-back approach and are what we call stalk-and-ambush hunters. I would never suggest that cats are lazy. Rather, they are efficient: this style of hunting is common among the feline species (the exception being cheetahs) and conserves a great deal of energy. Cats are not really in it for the marathon—which can be a boon when we are trying to play with them but have limited time to do so.

Stalk-and-ambush hunting often relies on an element of surprise. Predators sit and wait, and wait, and wait. In some cases, cats will travel to destinations that have proven successful for hunting in the past. This might be somewhere where there is abundant prey, such as an area of mouse burrows, bird feeders, and the like. With your own cat, you may notice them loitering by the toy closet when they feel like playing.

Innately attracted by sounds that suggest prey, such as rustling, scratching, or squeaking, cats will move toward the source of the sound to inspect further. They are highly motivated to chase any small object moving rapidly away from them or along a horizontal plane, telltale signs of a helpless, possibly tasty, and definitely frightened animal. Once the perfect location is found or signs of prey are detected, a cat may hunker down, often crouching in some grass or behind something that offers them a little bit of cover. Some cats seem to prefer ambushing out in the open, while other cats wait for rabbits or mice to pop out of burrows or warrens. Regardless of their preference before pouncing, the key is patience.

♥ ♥ ♥

This isn't a *Fatal Attraction* type of stalking; this is all about survival. When movement is observed or prey is heard (unlike dogs, cats use their sense of smell less often than sight and sound when seeking out prey), cats immediately stop what they are doing and flatten themselves to the ground. In what may best be described as an army crawl, cats will gradually approach their prey, staying low to avoid detection. Anyone with a cat has likely seen this behavior. They may alternate a crouched run with stillness, aside from the tip of the tail, which will often twitch in excitement. The head and neck are stretched out, the pupils are dilated, and the ears and whiskers are rotated forward, increasing the information available to all the cat's senses. At this point, cats will follow the tiniest movement of prey and make their way closer as they prepare for the final pounce. What works to a cat's advantage right now is their patience. They can wait. A long time.

It is a little hard to take our fuzzy house panthers seriously when right before attacking their prey they wiggle their butts, but nonetheless, this is what cats do. To be fair to them, this treading of the back paws is typically more exaggerated when cats are playing. If you have ever seen the butt wiggle, you know it is not the most dignified cat behavior—but it is definitely one of the cutest. The function of the butt wiggle is unclear; it could help propel the cat for the final pounce, or it may just be a way to relieve the muscle tension from holding still for so long.

Cats will decide at some point to dart forward, either low to the ground or in a jump, but generally with the back feet planted. Keeping the back feet touching the ground provides stability and the ability to make quick last-minute adjustments, including opting to run away if the prey is more formidable than originally thought.

Mice can be taken out with a one-paw smackdown. Grabbing prey that can fly (bugs, birds) typically requires both paws. Paws may also be used as a bludgeon when cats don't hit the mark on the first attempt. And if the front paws are not sufficient, cats will flop over and begin kicking prey with their back paws, further weakening their victim until they deliver the killing bite swiftly and effectively.

SENSES AT WORK

Like us, cats have five senses: vision, smell, taste, hearing, and touch. Some even consider cats' sense of balance to be a sixth sense. Let's look at how each of these senses chips in to help a cat function as a powerful hunting machine.

HOW EACH OF THE SENSES CONTRIBUTES TO HUNTING

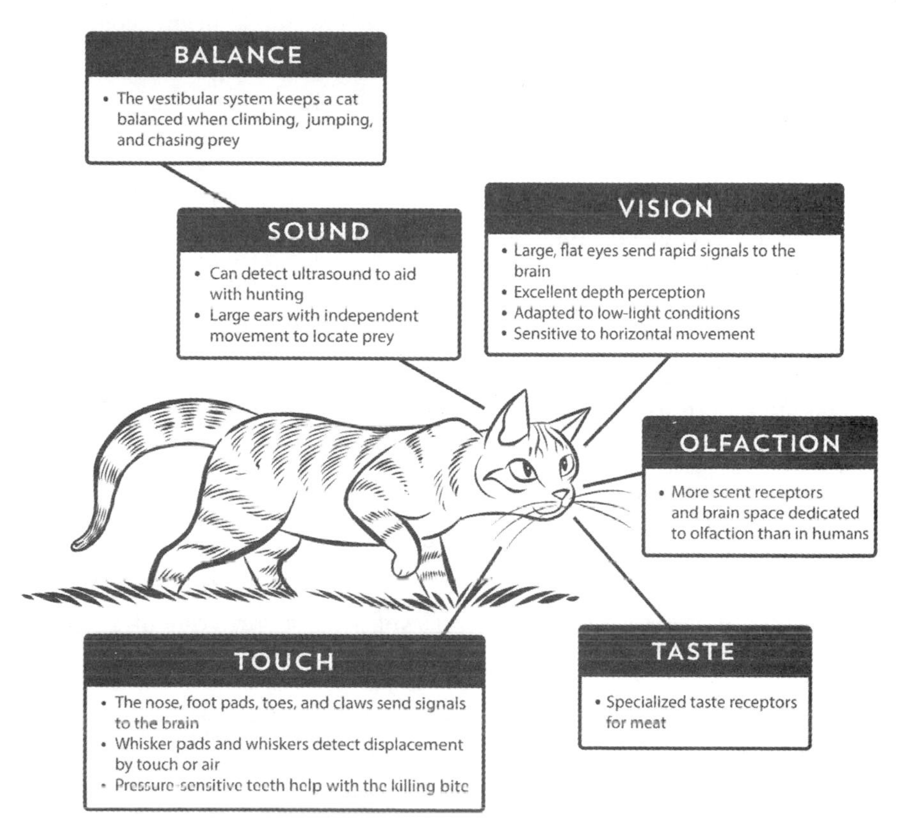

A VIEW TO A KILL

One of the many reasons we find cats so appealing (i.e., cute) is their large eyes, which invoke a sense of caregiving and affection in humans (this is a phenomenon known as *Kindchenschema*, or baby schema, where we respond to things we find cute with an urge to care for them). On the other hand, a mouse might contemplate those big eyes with trepidation, knowing that, much like the big bad wolf, a cat's response would be "all the better to see you with."

A cat's eyes are large in relation to their head, but also relatively flat compared to some other animals', providing for quicker signals from the pupil to the retina (the cells at the back of the eye) and hence, the brain, allowing for lightning-fast reflexes to pursue detected prey. Cats' eyes are especially sensitive to small movements, and their pupils are very flexible—opening almost to complete saucers at nighttime and closing to slits in the daytime to adjust to changes in light.

Their forward-facing eyes give them a wide range of vision, approximately two hundred degrees, which is a bit larger than what humans have and perfect for spotting a bird out of the corner of one's eye. One hundred degrees of their visual range is binocular vision (meaning overlapping vision between both eyes), which provides cats with excellent depth perception and the ability to detect prey moving across the horizon. But not *too* far across the horizon, as indoor cats are particularly shortsighted, with optimal focus when objects are two to six meters away (outdoor cats can see farther by comparison). Any object at a distance greater than that is going to be a bit fuzzy compared to what we can see. Ironically, cats also can't focus on anything that is too close either, so objects that are closer than around nine inches will be difficult for cats to see clearly.

YOUR TURN!

Start a play session with your cat by bringing the toy into their ideal range of vision (six to eighteen feet away), gradually moving it closer to them once you've captured their attention. Cats are most responsive to horizontal movements, but experiment as you wiggle the toy, varying distance and direction to see what your cat responds to best.

Despite apparently needing bifocals, cats' eyes are structured to give them an advantage when hunting—to see their prey when it is active. Rodents are out and about at dawn and dusk, meaning that to hunt them successfully, cats need to see well in low-light conditions. Cats' eyes are packed with a type of cell called *rods*, which magnify the effects of available light, allowing for the detection of movement when there isn't much light available.

Although we might stumble around in the dark, cats' eyes can take advantage of even small amounts of light to see shapes and movements with ease. To further amplify the effects of any available light, the backs of cats' eyes are lined with something called the *tapetum lucidum*, a collection of cells beneath the retina that light bounces off. These cells increase the amount of light detected by the rods and create the eyeshine you may have observed when taking photographs of your cat with a flash (but you can relax; your cat's glowing eyes are not a sign of demonic possession).

SMELLING DOUBLE

Cats live in an olfactory world we cannot imagine. They have a much stronger sense of smell than we do, with more receptor cells in the nose and more space in the brain dedicated to scent. Their excellent sense of smell is augmented by a secondary olfactory organ called the *vomeronasal organ* (VNO), which detects species-specific chemical signals called *pheromones*. Cats produce pheromones in scent glands that they have all over their bodies, including their cheeks, forehead, paws, flanks, and anal glands.

The VNO is in the roof of a cat's mouth. You may observe your cat carefully investigating areas that another cat has rubbed on or otherwise marked, opening their mouth slightly to take in more of the pheromone signal (this expression, which resembles a grimace, is known as *flehmen*). In this way, cats can send and receive messages from afar, using something akin to invisible ink as far as humans are concerned (also frequently described as "pee-mail"). Scent is a safe and secure way to send a message, and scent is social (but in a socially distanced kind of way).

When it comes to hunting, cats use their regular olfactory abilities (at least in a broad sense) to locate prey. Rodents also use urine marking to communicate with each other, and cats can eavesdrop on these signals to figure out where the mice are congregating. However, unlike dogs, cats likely do not use smell to track their prey for long distances. Most rodent urine marks are old news by the time the cat detects them, and so cats are more reliant on vision and hearing to detect a potential kill.

A TASTE FOR FLESH

All cats lack the amino acid sequence that codes for a gene called *TAS1R2*, which is one of the genes responsible for the ability to detect sweetness. As a result, it is thought that cats do not have the ability to taste sweetness in the way that we do. What can they taste instead? Meat. To further support cats' role as a super predator and carnivore, their tongues are equipped with receptors that detect ATP (adenosine triphosphate), a chemical compound that provides energy to cells and that is considered a physical signal for meat. If you're wondering what your cat is craving when they're hungry, you can likely narrow down the choices to . . . meat.

HEARING THINGS

We hear sounds when air vibrations travel through the ear canal and make contact with the eardrum, sending a signal to the brain that allows us to further distinguish features such as frequency (how high or low the pitch is) and volume (the intensity of the sound waves). Cats have one of the broadest ranges of hearing among mammals. They detect sounds that are higher (ultrasound) than what our human ears can detect, while also detecting sounds as low as humans can hear. Sensitivity to ultrasound likely allows cats to eavesdrop on rodent activity.

The structure of cats' ears further amplifies sound. The pinnae (the external part of the ear) are relatively large, with thirty-two muscles in each ear that allow for a wide range of movement and even the independent movement of each ear. You have likely watched your cat move and swivel their ears in response to a surprising noise. By moving their ears, cats can quickly and very accurately narrow down the direction and height of a sound source, guiding future steps toward capturing prey.

YOUR TURN!

Rustling sounds are *very* attractive to cats—try moving a toy underneath some tissue paper or a paper bag to get your cat's attention.

GETTING ON THEIR NERVES: MECHANORECEPTION GUIDES THE HUNT

We humans can understand the experience of touch. We have receptors in our skin that send information to the brain about pain, pressure, and temperature. Signals travel through the spinal cord of the central nervous system to specific areas of the brain, where the information may trigger action (such as pulling your hand away after touching a hot stove).

Like humans, cats are equipped with receptors sensitive to pain, pressure, and temperature. Many of these receptors are connected to hair follicles, and others are more sensitive to pressure on the skin (such as the sensation on your feet when you walk across a pebbly beach). When it comes to sensitivity, not all of the cat's body is created equally. Their nose, foot pads, toes, and even claws contain nerves that send signals to the brain. Their face and paws are incredibly sensitive to touch, and the signal is boosted by a hefty supply of whiskers.

Look at your cat's face. Take a really close look. Sometimes it's easy to overlook the seriousness of a cat's whiskers. But with careful inspection you'll see these stiff, thicker hairs on either side of your cat's mouth, their cheeks, and even their eyebrows and in front of their ears. On either side of the cat's mouth, between their lips and their nose, are what are called the *mystacial* (think "mustache") *whisker pads*. The whisker pads and the bases of all whiskers are where information really flows. Whiskers can help cats navigate through confined or dark spaces. Perhaps even more importantly, whiskers can make up for the fact that cats cannot see that well directly in front of their faces.

When a cat is pawing at or slapping prey, or preparing for the killing bite, they rely much more on their sense of touch than sight. (This can help explain why your cat may have difficulty locating a toy or treat that is *right. in. front. of. their. face.*)

Whiskers are sensitive enough to allow a cat to detect the degree, direction, and speed of whisker deflection, whether from direct touch or displacement by air. When whiskers are touched or displaced, cats have a reflex that turns their head in the direction of the touch. A touch of the lip can stimulate a cat to snap their jaws. These automatic responses increase the chance that a cat's hunting attempts will be successful.

Cats can also move their whiskers like a satellite dish focusing a signal. You may notice your cat pull their whiskers back when they feel threatened, or move their whiskers forward when they are excited. They can actually move their whiskers so far forward to almost wrap around a mouse!

Cats' paws are also equipped with several whiskers and hairs, and in the words of the esteemed cat behaviorist Dr. John Bradshaw, paws can be considered a "sixth sense organ." The exquisite sensitivity of the nerve endings in all those paw whiskers and hairs allows cats to fish for prey from a hole or crevice, and also lets them scoop that prey right into their mouths. A touch to the paw can lead to a strong reflexive movement, much like touching their face or lips.

Not to be outdone by whiskers, the claws and teeth provide further information. Movement of the claws gives cats signals about how much a prey animal is struggling under their grasp. Teeth are pressure sensitive and help cats determine the perfect location for the killing bite.

YOUR TURN!

Be sure to let your cat have contact with a toy—they may want to chew it, grasp it, or rub it on their face, but in general, they want the satisfaction of catching it! Make sure they are a successful hunter during your play session (even if you have to help them out a little bit).

AND TO BALANCE OUT THE SENSES . . . BALANCE

Cats are well-known for their exceptional grace, and like us, they have a vestibular system that helps keep them upright. This system detects the position of the head and its movement (including direction and speed) and then gives our brains feedback that allows us to do neat things like, for example, focus on this text while you shake your head left to right. The vestibular system, which is located in the middle and inner ear, controls our sense of balance.

The vestibular system and vision collaborate, allowing cats to focus on prey while moving in to attack. Add to that a dash of lickety-split reflexes, one extremely flexible spine, incredibly powerful back legs, and a long tail that can serve as a counterbalance when needed—now you've got yourself a stealth predator who can leap six times their height and who can pirouette, flip, and pounce with laser-sharp accuracy; they'll land on their feet every single time, eviscerate a mouse along the way, and move on to the next kill like it was nothing.

DO CATS PLAY WITH THEIR PREY?

Cats are often mischaracterized as cruel torturers because they appear to play with their prey. They may extend the hunt by batting at or tossing a mouse or releasing them from their grasp before pouncing on them again. While I have no doubt that cats do experience many good feelings during the hunting process (more on this later), many of the behaviors that appear playful to us are merely the cat being tentative about how successfully they can complete the job at hand.

Much like the bob and weave can protect boxers from an incoming punch, head bobbing in cats (which looks very silly to us) is a great way to avoid a bite or swipe to the face when taking on a large rodent. Batting at prey gently can give them feedback on how ready the prey is to respond aggressively to an attack. These tentative movements allow cats to test the strength and ability of their victim, making sure that the effort is worthwhile.

Cats may also experience some post-hunting joy, continuing to toss, bat, and mouth the dead animal; they may even leap and "dance." Dr. Paul Leyhausen has described this behavior as "play of relief," suggesting it allows for a release of built-up tension from a stressful hunt.

HUNTING MAKES CATS FEEL GOOD

There are reasons that, to many of us, behaviors such as sex and eating feel good. It is no coincidence that the things that help our species endure also give us pleasure—it ensures that we are motivated to do our best to both survive and create offspring! Our cats are no different. Although we can't ask them why they hunt, we can infer that their brains and bodies think it's a good idea. They are still motivated to hunt, even when not hungry. Hunting could even be described as a behavioral *need*. We can provide for that need by allowing cats to hunt toys!

There are chemicals in the mammalian brain (such as dopamine and serotonin) that are related to pleasure and mood, and it's believed that aspects of hunting and play can activate these chemicals. Without getting too deep into the neuroscience weeds, what we can say is that evidence suggests hunting is likely a feel-good activity for cats, from the stalking to the full tummy and everything in between.

POSTMORTEM MEAL

Once prey has been killed, it is time to eat. Cats may carry their prey (usually by the nape of the neck) to a location that offers cover and where they feel safer. House cats bring about a quarter of what they kill all the way back to their home turf (as many owners of indoor-outdoor cats who find "presents" on their doorstep will attest). You may even see similar behavior when you play with your cat, finding yourself in a tug-of-war as they try to carry a wand toy back to their favorite bed.

Inside the house, we also see prey-relocation behavior when cats remove some of their food from their bowl and carry it a short distance away before eating it off the floor. Moving the food dish to a space away from other pets, or to where the cat has more of a vantage point, can nip this messy habit in the bud.

Some cats may take a break before eating (killing is hard work) or will play with the dead animal before diving in. But other cats will continue hunting—a bird in the paws is worth at least two in the tree. When you don't know when and where your next meal is coming from, you can't always afford to take a lunch break. But the motivation to kill goes beyond just hunger. To that point, a recent study observed that cats walked away from almost half of what they killed. The instinct to hunt is very strong, and although strengthened by hunger, a cat need not be hungry to be motivated to hunt.

Dining behavior depends on the prey being consumed. Birds will be at least partially plucked before they're eaten; feathers can be pulled out with the teeth, or cats may lift the prey and shake their heads to loosen feathers. You can sometimes observe head-shaking behavior in your cats when they eat or play, so this behavior seems to be retained regardless of a cat's current hunting status, and it is unknown if the head shake might serve other functions (such as stunning the prey or snapping its neck).

WHY DOES MY CAT BRING ME PRESENTS?

Cats often prefer to eat in a more comfortable location than where they killed their prey. It may just be a coincidence that the comfortable place happens to be where you live too. (Take it as a compliment!) But the compliments end there: your pet isn't generously giving you a gift, nor are they making sure you don't starve. More likely than not, your cat might not be hungry enough to eat what they hunted. (After all, you are a reliable restaurant for your cat.) Of course, mother cats do bring prey to their kittens—so isn't it possible your cat is displaying a caretaking behavior toward you? While I can't offer a definitive yes or no, this explanation wears a bit thin when we consider that male cats will bring dead animals home at the same rate as female cats, yet male cats don't provide any care for their kittens.

If you're sick of waking up to dead birds in your slippers, there are actions you can take to reduce your cat's hunting behavior. Don't bother with a bell on your cat's collar, as evidence suggests bells are useless for preventing predation. A better option is the anti-hunting bib, which is a neoprene shield that attaches to a safety collar and hangs down in front of your cat's chest. They come in many adorable, brightly colored patterns.

Does your cat look ridiculous? Yes.

Is your cat mad at you for making them wear it? Probably.

Does your cat get used to it quickly? Yes.

Does the bib stop cats from killing birds and other animals? Surprisingly, multiple studies say *yes*! A 2007 study found that the bib stopped 81 percent of known hunters (of the feline kind) from nabbing birds, without interfering with other natural behaviors (such as jumping and running).

And if you don't want your cat to face the indignity of a bib, a recent study suggested that cats who are well fed *and* who get regular interactive playtime may be less inclined to hunt.

HUNTING IS PROHIBITED! NOW WHAT?

To kill is a delicate dance between predator and prey, and nowadays we have removed much of this drama from our cats' lives by keeping them primarily indoors and feeding them.

Even though most owners do not appreciate viewing their cat in this light, cats are natural-born killers. Predation has shaped every aspect of cats' evolution—every muscle in their bodies, all their senses, their lifestyle, their instincts, all shaped by the one thing they were born to do: hunt. Being successful hunters has allowed cats to thrive on almost every continent of the world.

And although we coexist with cats and throw food in a bowl for them, domestication has done very little to change their instincts to hunt. Hunting gives them life and it gives them pleasure.

We took hunting away from our cats, but we can give it back—by playing with them!

CHAPTER 3

From Prey to Play

HUNTING IN THE SAFETY OF YOUR OWN HOME

For cats, many aspects of hunting are mirrored in how they behave toward toys. In fact, we can stimulate all those hunting instincts without any mice or birds having to die—by providing our cats with interactive play.

The key here is *interactive*. You move the toy as if it were a bird, bug, or rodent, and your cat responds as if it were a bird, bug, or rodent. This is a lot different from leaving a few fuzzy mice and Ping-Pong balls lying around your house (maybe gathering dust). Although we'll take a deeper dive into the benefits of play and how to tailor interactive play to your cat's individual needs later in the book, let's start with the building blocks of a feline play session.

"BUT MY CAT DOESN'T PLAY"

I can't tell you how many times I have heard these exact words, because almost every one of my clients has said them to me. My experience is that humans don't offer enough play for their cats, they don't always know what to do to entice their cat to play, and they often give up easily. Feeling frustrated, these cat owners stop even trying to play with their cats at all, deciding they "just don't play."

Every cat has the hunting blueprint in their DNA, even if it's not always expressed in the same way for all kitties. With *every single one* of these clients who designated their cat to the "does not play" column, I have been able to demonstrate that their cat will in fact play (it's one of my secret power moves). It's always the best feeling to see their surprise when their cat bats at and pounces on a toy. Giving cat lovers techniques to get their cat (and them) off the couch is one of the reasons I wanted to get this book out into the world! Now, let's dive into some of those techniques.

WHEN IS THE BEST TIME TO PLAY WITH YOUR CAT?

The best time to play is *any time you are willing to pick up a toy and interact with your cat.* That said, making some adjustments based on your cat's naturally active periods may make things easier for the both of you.

Start by collecting some data. What time of day does your cat tend to romp or get the "zoomies"? Track their behavior for a week to look for clear patterns. An activity tracker designed for pets can give you very detailed data on exactly when your cat is most active.

Cats are crepuscular, meaning most active at dawn and dusk, so you can tap into their natural activity cycle and schedule play sessions during early morning or as the sun goes down. Excitement about food leads cats to display small bursts of activity before scheduled meals (called *FAA*—food anticipatory activity). Since your cat is already excited, before their regular mealtimes can be the perfect time to schedule play sessions. Cats play (and hunt) harder on an empty stomach, and play followed by a meal can help your cat settle down for a grooming session and a nap. But each cat is unique, and you may notice other times that your cat is alert or frisky—another opportunity to sneak in a quick play session.

Start with a framework you can stick to and build around that—ideally one to three times per day, depending on your cat's age and energy. Pick a time of day you can play with your cat consistently—all cats benefit from routine and predictability (as I'll discuss later in the book), and part of effective habit building is doing something every day at around the same time. You should still always be open to a little spontaneity. I happen to work from home, and sometimes when I'm taking a stretch break, that's a perfect opportunity for a short three-minute session with the feather wand and one of my cats.

HOW I PLAY WITH MY CATS

When my current batch of cats were kittens, we had daily play at around 6 a.m., 7 p.m., and 9 p.m., with the occasional afternoon session when time allowed. Each session was twenty to thirty minutes long. We focused a lot on the evening play because I wanted them to sleep through the night, but they were also pretty wild right when I woke up in the morning (I'm an early bird), so a preemptive play session to start off the day was the best way to stop my toes from getting tackled.

Now they are young adults, and our evening routine is that we always play at 7 p.m. for at least thirty minutes, with one or two supplementary and spontaneous shorter sessions (usually five to ten minutes) during the day. My cats also spend most of their daylight hours amusing themselves in our catio, which has taken some pressure off providing as much daytime play with us.

HOW TO MAKE PLAYING WITH YOUR CAT A DAILY HABIT

- Remember, you are doing this for your cat's health and well being. It is just as important as food, water, and a clean litter box.
- Start small the goal is to have at least one play session happen on a schedule every day.
- Make it fun for both of you.

- Go public—be accountable by sharing pics or videos on social media.
- Set a schedule and stick to it—play *every* day at around the same time(s).
- Make play easy—organize your cat's toys so you can access them and switch them out quickly.
- Make it part of your existing routine, e.g., play after you brush your teeth or before you eat dinner.
- Reward yourself and your cat at the end! We always like to say "good sesh" to our cats to celebrate the success. But why not end with a treat?
- Focus! Try not to multitask when playing with your cat.

"HOW LONG DO I HAVE TO PLAY?"

People always want to know a specific number of minutes they should play with their cat (also, they often frame this question as "How long *do I have to* play with my cat?" rather than "How long *can I* play with my cat?"). It's not like baking a cake (forty-five minutes and they're done).

For kittens up to one year, it's not unrealistic to think they benefit from up to ninety minutes of play most days. That doesn't necessarily mean that you are waving a wand all day (I mean, it could . . .), but it does mean that you are going to have to put in a lot of time and effort that first year. You can save some time by:

- Making sure kittens have age-appropriate playmates (i.e., other kittens)
- Judiciously using solo and automated toys (see chapter 13)
- Having a catio where your cats can spend safe time outdoors, watching birds and squirrels
- Utilizing other sensory enrichment for your cat, such as window perches, climbers, food puzzles, and bird feeders to watch

But none of these things will completely replace you playing with your kitten at least three times a day, for fifteen to thirty minutes at each stretch.

As your kitten matures, they will settle down and most will need much less play than they did when they were young. For a young adult cat, you may find that one or two long play sessions a day works well, or you may try three or four short sessions, depending on their stamina and interest, and of course your daily schedule. For older or mature cats, you may see the play sessions are much shorter (five to ten minutes), and your cat may only need one or two play bouts each day to be happy. But every cat is an individual, and some are more playful than others, and that will be part of your equation when setting up your cat's play schedule.

BUILD IN SOME BUFFER TIME

Warm-up time is another consideration. Some cats are ready to play as soon as you open the toy closet. Other cats need quite a bit of "warm-up"—meaning, they may watch the toy with limited interest for several minutes before they decide to actually make a move. Or they might move quite slowly or lazily for a few minutes before the chasing or jumping begins. If your cat isn't racing right out of the gate, don't get discouraged (as I like to say, "Playing while laying is still playing!"). Sometimes you must invest a little time in play before giving up, or try switching to a new toy to see if that increases interest. Keep in mind that cats hunt in short bursts, with frequent encounters with prey. Before they initiate play, cats may also need auditory or visual stimulation to get warmed up, just like hearing or seeing prey would instigate their hunting instincts.

THE GOLDEN RULES OF PLAY

When it comes to cat play, I'm mostly an anything-goes type—give it a try and see if your cat likes it. But I do have a few golden rules—the principles all play sessions must adhere to first and foremost.

- **Don't scare your cat**. Part of play is helping your cat feel like a brave, competent hunter. You must be sensitive to your sensitive cat. Some cats really don't like larger prey, loud noises, or sudden movements, so watch your cat's body language carefully and adjust your toy or movements as needed. I'll share more about scaredy-cats in chapter 11.

- **Safety first**. It goes without saying—but keep safety in mind, and also keep your cat's physical limitations in mind! We don't want any sports injuries, we don't want any cats getting tangled in a string, and we don't want your cat swallowing anything that can lead to a nasty intestinal obstruction.

- **Don't irritate your cat**. This may be the least intuitive of the golden rules of play. I've found that a lot of cat folks have a hard time discriminating between irritation and engagement. Case in point: I observe a lot of people touching their cat with a wand toy in an attempt to play. Please don't poke your cat with the toy. No self-respecting bird or mouse would walk up to your cat and get right in their face or tickle them! People often make this move, I suspect, because it tends to get a reaction from the cat (usually an irritated one rather than a playful one), which they mistake for interest.

- **Let them win**. Another thing that can be annoying to cats is never getting to touch the toy. If you always wave the toy just out of your cat's reach, or you only use a laser pointer for play, your cat can never complete the hunting cycle of catching the prey—which makes play more frustrating than fun. Instead, allow your cat several opportunities to touch the toy, even if the toy quickly escapes! I like to end a session with some feline success, where they get to hold and bite the toy basically for as long as they want (or until you need to wrap things up)

- **No roughhousing**. Please, never wrestle with your cat or use your hand as a toy. This is confusing to your cat and sends them mixed signals—sometimes body parts are for biting and kicking, but other times they are for petting and cuddles. Roughhousing can also make your cat defensive about other handling, such as hands approaching their face. I've seen more than a few cats develop aggression problems due to being roughhoused with, and even if it doesn't bother you, it is likely stressful to your cat. You may also live with other people who do *not* appreciate that your cat has been trained to bite hands (or other body parts). If you're playing with your cat, there should always be a toy in your hand, and your cat's focus should always be on the toy, not your body.

THE STRUCTURE OF A SESSION

For most cats, you can think of the play session much like you would an exercise session for yourself: there's a warm-up, the high-intensity part, then the cooldown. But even within that framework, we know there are lots of ways to exercise if you are human—you can run for thirty minutes straight, or you can alternate five minutes of running with five minutes of walking until you have reached your time or distance goal. Your pace when playing with your cat will depend a lot on your cat—their age, their play style, and how long it takes them to warm up.

For many cats, the warm-up is the most important. In fact, it may be the destination. I jokingly call this part *kitty foreplay*, as you are really taking your time and enjoying the process of getting your cat to notice the toy, then focus on the toy, hopefully increasing your cat's engagement and enjoyment as the play session continues. But as I'll discuss later, in chapter 12, about playing with senior cats, the warm-up is often the entirety of the play, and that's just fine.

The warm-up is you introducing the toy to your cat. Remember that all your cat's senses can be part of this process. This may start with you walking toward the toy closet, which may generate excitement in your cat. Or you may be introducing a sound, such as the rustling of a feather against the couch leg, or playing some bird sounds in the background. Or it could be the visual stimulation of a toy slowly moving into view. Sometimes I even get my cats in the mood with a little bump of catnip or silvervine.

On the other hand, young cats can go, go, go—often until they are panting or overexcited. You may have to watch them and give breaks accordingly or keep the pace a bit on the slower and steadier side. Especially when the weather is hot, or if you have a cat with heart disease, you want to be aware of not going so hard and so fast that they can barely breathe. I've also seen some cats become so excited that they had a hard time calming down *behaviorally*, which can be problematic if you have other cats in the home. If the toy is put away too soon, another cat in the house quickly looks like a nice mouse substitute, and chasing or even fighting can result.

Many cats appreciate a burst of excitement followed by a lull or even a break. Some cats will do best when you repeat the warm-up/high-intensity/cooldown cycle a few times, with each cycle becoming shorter as you revisit it. This makes sense when you think of cats' stalk-and-rush hunting style. Alternating the pace can work well if you have multiple cats and not enough humans to split up the play duties. Especially if the cats get along and can spend some time watching each other, you can sometimes just alternate who is the "star" of the play, switching to another focal cat when one is ready for a break. You might cycle through each cat two or three times and then everyone is happy.

THE TEMPTATION TO BE LAZY

I can't emphasize enough how important it is to be engaged in the play session. Ever played a game or tried to have a conversation with someone who was checking their phone every two minutes? Were they giving it their all? I doubt it.

I've been there. I've thought to myself, "Crap, it's time to play with the cat." I've picked up the toy, carried it back to the couch, and sat my ass down while watching TV, moving the toy just off to the side halfheartedly. But really I was disengaged, watching TV, not playing with my cat. Now, at first, my cat obliged, and batted at the toy (at least I think she did; I wasn't really watching her that carefully). One night, as I walked toward the toy closet, I saw her eyes light up with anticipation of playtime. And then I *swear* I saw the disappointment on her face as I marched myself back to the couch. She was bored of my half-assed attempt to play in the same damn place night after night. That look on her face gave me a sense of guilt and a renewed commitment to pay attention when I play with my cats!

My advice to you is: Don't multitask when you are playing.

The play should be the focus. And if you are physically able, incorporate some movement and new locations into the play, changing things up to the extent you can.

TOP FIVE TIPS FOR PLAY

1. Act like prey.
2. Move the toy slower. Slower. Maybe even a little slower.
3. Vary your movements!
4. When all else fails, use the stick end of the toy, moving it under a towel or rug.
5. Make sure the cat is successful in their hunt and has time to handle and interact with their prey (however they prefer to do so!).

MOVING THE TOY: SPECIFIC TECHNIQUES FOR SUCCESS

I had to think long and hard about this section because it's the sort of thing that can be much easier to show than tell. A play session can be a song and dance that you perform for your cat's amusement, or it can be a true tango between you and your cat (plus the toy). You will learn to make the slightest adjustments as you look for signs in your cats—the focused stare, the dilated pupils, the alert whiskers, the butt wiggle—that all let you know your cat is feeling it. There will be perfect times to pick up the pace and swoop the toy out of reach, and also perfect times to go slow or pause altogether. I have a few "moves" that I return to again and again during play. I'm going to do my best to describe them to you now.

GETTING STARTED

- **The dramatic introduction**: In this case, I might immediately start the play session with some large movements of the toy, usually up high in the air. I always do so several feet away from my cats and only if they are awake, so they aren't startled. Think of a bird swooping overhead. A dramatic introduction could also be conducted in a separate room from your cat, so their curiosity is piqued by the sound of the toy moving in the other room. However, they should take notice quickly. If your cat does not come to inspect, slow down the movement a bit and swoop a little more gently into whatever room they are in.

- **The slow onset**: This is kind of the opposite of the dramatic introduction. In this case, I'm tapping into my cat's amazing abilities to detect the smallest movements or rustling sounds to get their hunting juices flowing. A slow onset can happen anywhere—the floor, on a piece of furniture, under a rug, or by just barely tapping the stick of a toy against a chair or table leg. In this case, you're letting your cat slowly detect that the play session has begun.

FOR THE BIRDS

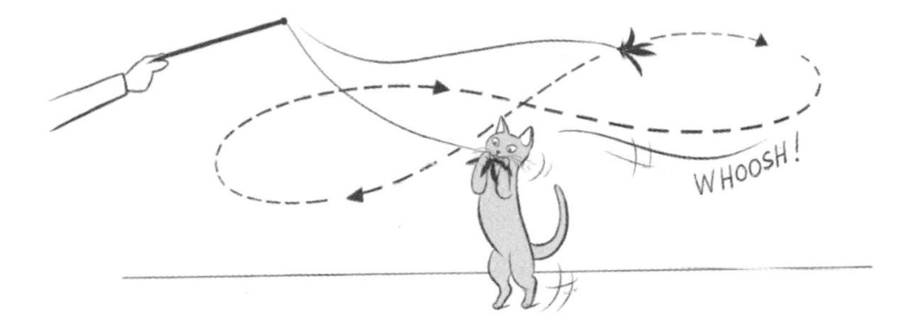

- **Flutter and wow:** This move is best with a feather wand such as the Da Bird (or similar—basically any hefty feather on a heavy string and sturdy wand). The real appeal of these toys is their ability to make realistic sounds as they move through the air, and how efficiently they can "fly" through a large space with the smallest of movements of your wrist. Imagine the bird has entered your home and is somewhere between a proud hawk surveying the land below and a panicked bird that has gotten itself into an indoor space and now wants out. You can incorporate some figure-eight motions into your swooping as well.

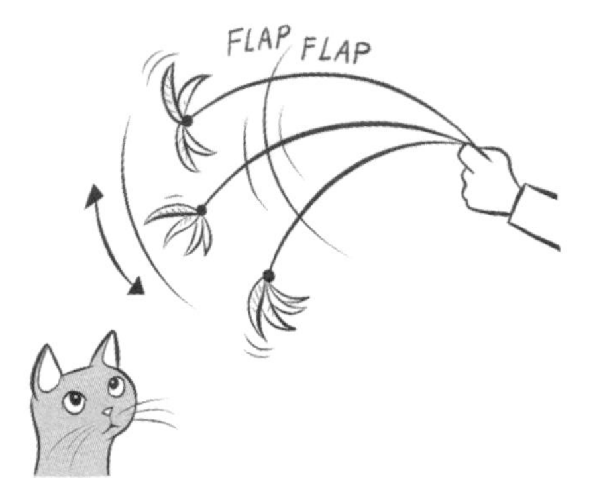

- **Winging it**: We're still in the bird family, but this move is for the classic plastic wand with a collection of feathers at the end. Depending on how long the wand part is and where you hold the wand (i.e., how close to the feathers), you'll be able to control the amount of movement of the bird part more tightly or allow for more free movement (we'll get into the physics of cat play in the next chapter). Vary the speed of the movement to see what gets your cat's interest—fast movement creates more sound (flapping), which can be exciting for some cats and a little scary for others.

- **Overhead:** Although we eventually bring things down to our cat's level, remember that they spend a lot of time watching prey overhead, whether flying in the sky or leaping through trees. Fly your bird toy up near the ceiling or to the highest cat-accessible perches you have, such as a cat tree, ideally landing somewhere that your cat can also climb to.

- **The grounded bird**: What goes up must come down, and at some point, it's time for your bird to land. Land two to six feet away from your cat, and then think about your next move. If your bird is foraging for seed, lift the toy so the feather is upright but touching the ground. You can then simulate the bobbing motion that a bird might make as they walk and peck around. On the other hand, perhaps your bird is wounded, hopping around a bit, or trying to fly but having some challenges getting off the ground. Your bird may even succeed in flying up to land on a low surface, such as a cat-tree platform or the couch.

- **The just-out-of-reach swoop for backflips**: If you have a cat who gets aerial when they play, there are a few ways to get them off the ground. If you're using Da Bird (which I highly recommend) or similar, you may find you don't have to do much to get your cat off the ground. I have observed that there is a sweet spot for some cats to get them to do a backflip, which involves moving the toy in an upside-down U-shaped arc over their head, moving the toy just out of your cat's reach at first. As they follow the movement visually and reach their paws out to touch the toy and simultaneously jump, this often propels them into a head-over-heels leap that would put some Olympians to shame.

GROUND PLAY

- **Mouse under the rug**: At the risk of making a broad generalization, I'm going to say it: Cats really love to watch a toy moving under something. It doesn't have to be a rug; it could be a blanket, a towel, a piece of tissue paper. Just move a toy under it and occasionally poke the toy out from the edge. That wiggling lump peeking out is almost always irresistible. When in doubt, try the stick end of the toy and then wonder why you tried all the other fancy moves and expensive toys.

- **Tiny, shivering-with-fear mouse**: Don't forget that you are pretending to be a prey animal who might be on the verge of being pounced on and torn open. In response to a threat, most animals turn to fight or flight. But there's a third *F*, and many small animals such as mice will freeze when they realize they have been detected by predators. However, that mouse can hold completely still for only so long before they get tired, or shiver, or try to sneak away (those small movements are what often get cats the *most* excited). Experiment with some small, jerky movements in between periods of stillness.

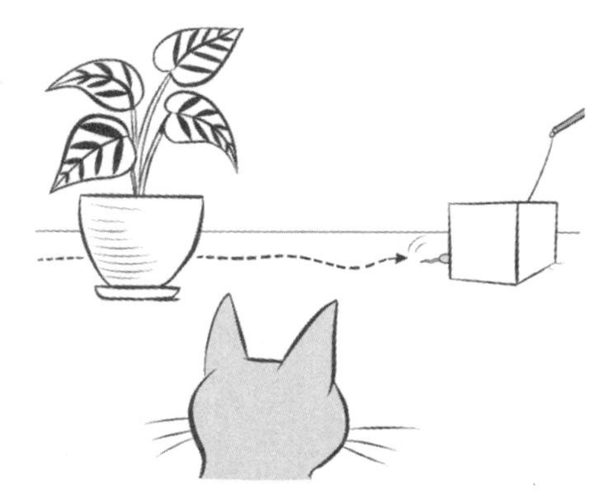

- **Sneaking around**: This is the prey toy that "runs" from one hiding spot to the next—from behind the cat food bowl to under the couch to behind a plant pot and then behind the curtains. You can even add more hiding spots (e.g., cardboard boxes, cat scratchers, furniture) to give your prey animal more opportunities to grab cover between trying to escape your cat's lethal jaws.

- **Around the corner and out of sight**: There's something about prey just out of reach that drives cats wild—and this is where you can use doorways and corners that exist in your home to sneak around. I especially love this move with a peacock feather—very slowly and gently moving the toy around an edge until it crosses into the threshold of invisibility, which is the point at which your cat *must know what happened.* You can also do this move with anything that leads a toy to be slowly obscured, such as under a door or into a paper tube (try moving a toy in and out of a poster tube for fun!).

- **Slithering snake**: If you've seen a snake, you know what to do. This works best with toys like the Cat Charmer or a shoelace. Stay low to the ground and move your hand in an S shape to create that snakelike movement. You can vary the speed based on what your cat responds to best.

- **The slowest crawl ever**: One thing that humans have a hard time doing is slowing down. Often my first piece of advice is just: "Try that again, but slower." The slowest crawl is just that: moving any toy very, very slowly, usually along a surface (whether horizontal or vertical).

MOVING THE TOY

If you have a Cat Dancer toy or similar—usually a wire rather than a string, with the lure attached to a handle—you have two options.

- **Barely in control**: Option one: You can hold the toy by the handle, letting the wire move as it will. This can create a slightly more kinetic, frenzied motion depending on the size and weight of the lure. The movements of the toy might look a bit like those of a mosquito or crane fly.

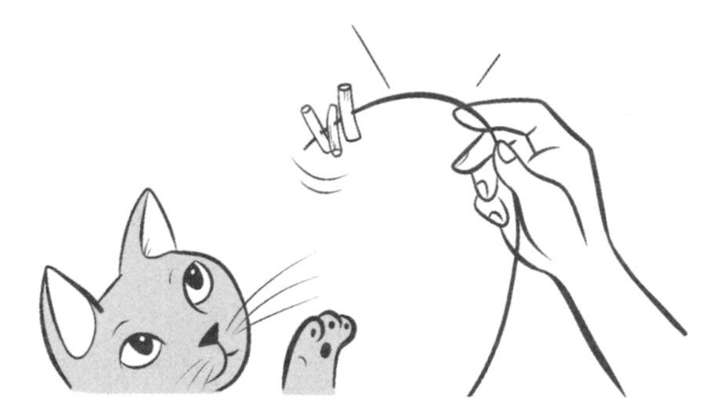

- **Under control**: Option two with that wire or otherwise floppy toy is to hold the wire close to the lure end. (I usually like to give a few inches' berth to make sure my hands are out of the way of my cat's teeth or claws.) This move allows you to have precise control over the toy's behavior, in a way that might resemble a fly buzzing or a bee visiting from flower to flower.

SMALL OBJECTS

- **Toy roll or toss:** Solo play toys such as balls and fuzzy mice aren't just for solo play. We often end a play session with a cascade of mice flying across the room. I'm also a sucker for small foam balls that look like soccer balls, and sparkly pom-poms (luckily my cats like them too). Roll them, toss them, skitter them across the floor, then pick them up and start all over again. I like toys that make a little bit of noise (like the mice that rattle or the crinkly balls) so I can get a cat's attention, and I find that throwing them at an arc just above your cat's head is a great way to get all four paws off the ground. Have fun with it, just move your toys thoughtfully and with your cat's interests in mind. The key here is that there is movement of the toy, at least long enough to hold your cat's interest. If you're especially lucky, you might have a cat who innately fetches or who you have trained to fetch. Otherwise, you will also get some exercise from this activity.

- **It's a bird, it's a plane, it's a Frisbee?** Much like rolling or tossing a toy, we can also use toys that glide. I've known cats who loved to chase yogurt lids when they were tossed much like you might fling a Frisbee. My cats have also enjoyed chasing paper airplanes.

NOT FEELING IT

If your cats seem sluggish or disinterested, don't give up yet. It's not unusual for me to try anywhere from two to five toys during a play session (and sometimes even more—this is why I like to keep my interactive toys handy and well organized, as I discuss in the next chapter). Try changing something else about the play—the location, the texture of the toy, or the speed of toy movements. Try moving a piece of furniture, such as a chair or cat tree, or placing a new cardboard box in the middle of the floor. On a tough night, I might introduce a robotic toy for a few minutes to pique my cats' interests, or even put a bird/mouse video on the TV. Sometimes I just need to get them off the couch!

But seriously, everyone has an off night every now and then. The important thing is to not give up. Instead—assess. If this is an ongoing pattern, then you might need to think about the time and place you are choosing to play. You might need to invest in buying or making some new toy lures. You might have to try some new movements. You might need to read (or reread) this book.

COOL IT NOW

Much like you want to bring down your heart rate slowly after you exercise, you can do the same for your cat. This is both a physical and mental cooldown. Start the cooldown by slowing down the movement of the toy. Your cat may still be interested at this point, and that is okay. Then start introducing longer and longer periods where the toy is completely still. The toy is dying a slow, lingering death that will gradually reduce your cat's excitement. Let the toy be completely still for a minute or two before "calling it." When you take the toy away, move it slowly, and if possible and safe to do so, cover the lure part with your hand so you are reducing their exposure to the exciting part (the prey-like stimulus).

By the time you end the session, your cat should be looking away from the toy, or perhaps they've moved on to a new behavior such as grooming. This is the perfect time to offer your cat a treat or snack—it ends the play session with a distracting reward, and it also replicates the natural cycle of hunting, which would end with a meal.

CAN'T STOP, WON'T STOP

Some cats can become quite possessive over a toy or might not be ready to end the session. If your cat still has a toy tightly secured in their jaws but it's time for play to end, what should you do? First of all, don't yank the toy out of your cat's mouth. That only increases the chance they will bite down even harder. They might break the toy and swallow something they probably shouldn't. Instead, get your cat to drop the toy of their own volition. There are a few ways you can do this:

1. Wait and hold the toy with the loosest grip you can to make sure your cat is safe but isn't tugging. As the toy "dies," your cat will lose interest and eventually let go.

2. With your free hand, move another toy a few feet away, and wait for your cat to let go of the current toy in pursuit of the new toy. Ideally the new toy is interesting but not exciting enough to simulate the grab-and-hold to the same extent. If they do, go back to idea number 1.

3. Skitter a small solo toy or hard treat away from the toy. Most cats will drop the toy in pursuit of the food snack. If that doesn't work, just place the treat a few feet away and wait for your cat to drop the toy.

4. You may also want to usher or lure your other cats out of the room and into another area of the house. Sometimes the competition of other cats around makes

cats hold on harder to their prey. On the other hand, cats with FOMO (fear of missing out) will drop everything as soon as they think the other cats are getting something better in the other room.

I don't recommend just letting your cat carry the toy endlessly around the house, mostly for safety reasons (we want to make sure that your cat doesn't try to eat the toy or get a string wrapped around any body parts). I prefer the human puts the toy away!

CHAPTER 4

The Right Toy for the Job

Let's talk toys.

Yes, there are many toys that will work just fine, but they are like how a card shark might feel, perhaps, playing Go Fish versus high-stakes poker. We want to bring your cat the high-stakes-poker version of play—using toys that will bring them to new depths of excitement and pure happiness. Remember, the most successful play is that which activates their hunting instincts.

For effective interactive cat play, you want to have the right tools. There is a huge array of cat toys out there. Some are great, some are terrible, and sometimes it's hard to tell the difference until you've tested them out with your cat.

Right now, we are going to focus on *interactive toys*. You might hear them referred to as fishing pole toys, wand toys, or teasers. These different names all refer to the same basic concept: a stick that you hold, with something attached at the end. They might have wires, strings, fabric, or similar material, from which a small object is hanging. At the end is usually some type of lure, perhaps bird feathers, a fabric mouse, or streamers. Poles and strings may be long or short, and lures may be big or small. I recommend having a selection of toys to choose from, aiming for at least five to ten different types of purchased toys, plus plenty of homemade options.

Now let's look at the parts of that toy, starting with the pole.

Choose a handle that is comfortable when you hold it. My personal favorite style is a wand with a padded handle, and a rod that is two to three feet long. The padded handle is more comfortable for those lengthy play sessions, and the longer rod allows you to play while standing up and walking around. It also allows you to reach high-up spaces, should your cat enjoy playing from the tippy top of a cat tree. That said, you may find that the perfect toy for your cat has a short wand. The rod itself can also be very stiff (e.g., a wooden dowel) or more flexible (plastic).

The pole may have a toy attached directly to the end. Many wands are just a plastic rod with some feathers, tinsel, or other small objects attached. Simple, elegant, effective. Some toys may also have a bell attached. Keep in mind that some cats will find the jingle of the bell a little scary. If that's the case, you can just remove the bell and keep the feathers.

Most toys will have some type of cord, wire, or string between the wand and the lure. Each of these attachments will give you a slightly different play experience. A wire will give you rigid movements that you can control more subtly. String or cord is good for large, sweeping movements of the toy. I really like toys made with monofilament fishing line, because they offer a bit of the benefits of both wire and string, with the bonus that the line is translucent. I suspect that this gives the lure more of the illusion of prey and less like "the thing attached to my parent's arm."

Take a good look at how the toy is made. Often what distinguishes a cheaply made toy from one I would consider worthwhile is how the string is attached to the rod and how the lure is attached to the string. Cats can get rough during play and may rip a string right out of the wand if it's poorly attached.

An eye hook drilled into the wand or a plastic cap covering the area where the string attaches to the wand adds durability to the toy. If the string is attached with just glue or you can observe cheaply made parts, the toy most likely will not last as long. A toy like this may provide your cat with some brief joy, but your wallet may not appreciate it.

Now we are at perhaps the most important part of the toy—the lure. This is, after all, the part that will offer (in most cases) the predatory appeal to your cat. The lure may be directly tied or glued to the attaching string, but most quality toys will use a barrel swivel (also used for fishing lures) that allows you to switch out lures. Having lure attachments that you can change gives you more bang for your play-wand buck, and many lures are sold independently of the wand, which can save you money while increasing your cat's excitement.

When choosing lures, think about what your cat might like to hunt. Look for lures that have properties similar to the animals they would naturally prey upon. Cats tend to prefer toys with fur

EYE HOOK OR CAP
WIRE, STRING, OR CORD
BARREL SWIVEL
HANDLE
LURE
ALTERNATIVE LURES
CAT CATCHER
DA BIRD
WORM
BUG
HOMEMADE LURES
OPTION: ATTACH LURE TO THE ROD DIRECTLY
WOODEN DOWEL WITH EYE HOOK
SMALL PIECE OF FABRIC
STRIPS OF KRAFT PAPER
A GREEN BEAN

or feathers, but those are not the only qualities that can be al*lur*ing—which is why I strongly recommend trying a variety of lures and rotating them regularly.

I lump cats' hunting preferences into four primary prey categories: mice, birds, bugs, and lizards and snakes. You want lures that resemble these prey animals in one way or another. Let's look at some features of each prey animal that can help inform our toy choices and play style.

MICE: Mice are small, furry, and have a long tail. They squeak and also make ultrasonic calls out of our range of hearing (but not out of our cat's range of hearing). If you watch videos of mice (or happen to have some mice in your environs), you will often see start-stop movements alternating with sprints. Mice are jittery. They can climb up vertical walls and do a little bit of jumping too. Mice like to nest in walls and crawl spaces and will also form burrows outdoors. They enter spaces through small holes and crevices, which cats are naturally attracted to. In fact, Dr. Leyhausen's studies of cats hunting determined that some cats might even be more interested in the holes and crevices that mice like to hide in than the mice themselves.

BIRDS: Birds can be large or small, but they've all got feathers and most of them tweet. Birds can walk, hop around, and peck, and then flutter away at a moment's notice. In the air, they can swoop, hover, or fly from one tree branch to another. Sometimes they're flapping, sometimes they're gliding. They might rest on the ground or on elevated surfaces.

BUGS: Bugs tend to be small and quiet. Their movements can be slow and sedentary or fast and frenetic. But the movements tend to appear a bit more random, alternating between circular swirling in the air and more zigzag movements. Bugs may fly, hover, land, freeze, or crawl. They appear to be attracted to windows, lights, and corners, and often nest or hide in crevices or behind curtains.

LIZARDS AND SNAKES: Most lizards crawl around on four legs, occasionally pausing to do some pushups (which male lizards do to attract females). Their tail drags behind them and may help them with crawling. Many lizards can freely climb vertical walls, due to special hairs on the bottoms of their feet. Snakes have no feet and instead must slither around. They use their belly muscles to glide and slide forward while folding up almost like an accordion.

If you're reading these descriptions, scratching your head, and thinking, "What . . . ?" I highly suggest watching videos made for feline entertainment. There are several available for free online that feature assorted prey animals. Carefully watch how the animals move and see if you can emulate their movements with a toy. If you and your cat have a watch party together, you might notice that certain things on the video get your cat's attention more easily than others. Those are the movements to become fluent in (or at least start with)!

PREY TO PLAY

Perhaps now you've narrowed down the prey preferences of your cat or want to try a few toys of each type. How to decide and make your final purchase?

MICE: Look for something small, mouse-shaped, fuzzy, and with a tail. Ideally, the rod has some flexibility and the tail of the mouse is long enough to be a bit creepy and realistic. Look for a rod and string that will allow you to drag the toy along the floor and also mimic hopping and climbing. (Note: I would also put rat- and squirrel-like toys in the "mouse" category; they're just a little bigger.)

MOUSELIKE

BIRDS: You will likely need more than one bird toy. I highly recommend getting a few wands with feathers attached with a string and swivel for the realistic fluttering movement and sounds that get cats excited. You'll also want one or two classic feather wands—just a stick with feathers on the end, no line necessary. Finally, consider a naturally molted peacock feather (some farms and zoos sell them, and you can find them readily online) to round out your bird collection.

FEATHERS

BUGS: When looking for insect lures, think small and creepy-crawly. My favorite is the classic Cat Dancer toy, which is a wire with small curls of cardboard at each end. It doesn't really look like a bug but has a classic slow fly kind of movement to it. There are also several options for bees, beetles, caterpillars, and dragonflies. Some toys will have wings or small legs attached.

BUGLIKE

LIZARDS AND SNAKES: One of my favorite snakelike toys is just a tie-dyed strip of fabric dangling from a rod. Simple, slithery. To give your cat a reptilian or amphibious experience, look for lures that are long and stringlike. (Bathrobe ties and shoelaces also make fabulous lizard or snake substitutes.)

SNAKELIKE

TOYS THAT DEFY LABELS: Don't let yourself be limited by these four categories (there are certainly other animals that cats will stalk and hunt, such as fish and rabbits). And sometimes toys break all the rules, but your cat still loves them. As humans we are so visual that we sometimes forget that there may be other features of a toy your cat will be just as, if not more, attracted to beyond what it looks like, such as the sound a toy makes when it moves, or how it feels under their paws or in their mouth.

THE ISLAND OF MISFIT TOYS

Do not feel like you must throw away your feather wand just because it's looking less plush than it used to. Sometimes my cat's favorite toy has been the feather wand with just one or two sad, scraggly feathers barely clinging to it.

THE EYES HAVE IT! DOES THE COLOR OF A TOY MATTER?

Humans with full-color vision see the vibrancy of a rainbow in our surroundings. But the same colors look a bit drab to cats; if your main interest in life is chasing brown mice, you do not need to discriminate fire-engine red from lime green. Cats are not color-blind; they can see colors in the blue-violet and yellow-green ends of the spectrum but with much less brilliance. Red colors appear mostly gray.

What does this mean for play? The color of the toy may be less important to our cats than it is to us! Many cat toys are designed to be appealing to humans (do cats really care that a toy looks like a lemon or slice of pizza? No, but I definitely do . . .), so what jumps out to us at the pet store or when we're shopping online may not be what your cat enjoys. However, since cats do see shades of blues and yellows, toys in those colors may stand out the most (hello, catnip lemon).

BIGGER ISN'T NECESSARILY BETTER

You might be tempted to get toys with a large lure on the end, but research has shown that cats are much more likely to kick, bite, and clutch a small toy compared to a large toy. In a 1998 study of cats playing with either a small mouse-sized toy or a larger rat-sized toy, cats were more likely to avoid the larger toy, and they even showed behavior suggesting they were a little afraid of it. Hunger made cats more likely to approach and sniff the large toy but was not enough to get them to play with it.

FEELINGS . . . NOTHING MORE THAN FEELINGS

Think about the textures a cat would experience while hunting. Their paws are exquisitely sensitive and equipped to detect different textures, from the sleek yet firm feathers of a bird, to the warm softness of a mouse's skin, to the dried husky feel of insect wings. When you choose lures or other cat toys, I encourage you to touch them and imagine what your cat will experience when they play with them.

As much as it grosses me out (I've been vegetarian for over thirty-five years), I have to say that for many cats, only real-fur mice will do (they're usually made of rabbit fur—I'm sorry, bunnies!). I don't know how they can sense the death in the toy, but for my cats, there's nothing like the real thing, and they will play more vigorously, fetching, chewing, and tossing those rabbit-fur mice in a way they just don't do with the faux-fur ones.

Another aspect of play we often don't consider is that when a cat is hunting, their prey changes during the kill. They will bleed and change body temperature. The body will be ripped open and lose form. Feathers may fall out, skin may be ripped off, wings may be shredded. Yes, it's a bit disgusting, but it is just a fact.

But cat toys are usually made to be sturdy! In fact, most of us would probably get irritated if a lure only survived one or two plays before it was destroyed. We paid good money for those toys! What's the happy medium here?

I recommend buying well-built toys, but also thinking about ways to provide your cat with an outlet for their appetite for destruction. This could include:

- Buying a few less-well-made toys, accepting that they might get destroyed
- Making homemade lures out of kraft paper (a heavier paper often used for packaging and shopping bags),

tissue, toilet paper, or cardboard (shreddable and recyclable!)

- You can even put a few pieces of dry food in a tissue paper "bindle" that you can tie to the end of a wand toy, allowing your cat to rip it open for treats

The most important thing if your cat is the seek-and-destroy type is to make sure they aren't ingesting any of those small bits that are ripped off a toy (such as beads, buttons, strings, or even feathers). If your cat ingests a few small pieces of paper, they will probably be okay. But paper isn't food, and you certainly don't want your cat snacking on it. And other items can cause problematic intestinal blockages.

Some cats have a condition called *pica*, which is when they chew and ingest non-food items. Pica can be dangerous, and for these kitties, the sturdier the toy, the better. If your cat has pica, you should work with your veterinarian and a qualified behavior professional to get to the root cause.

THE PHYSICS OF CAT PLAY

I recommend having a few different toy styles, because it allows you to experiment and provide your cat with some different play experiences. There's no right or wrong type of toy—you just have to know what to do with each type.

Although I thought I knew a lot about how to wiggle cat toys around, I wanted to check my intuition, so I chatted with professor, physicist, and cat-lover Dr. Greg Gbur. He lives with several cats and even wrote a book about cats and physics, so I knew this was the expert I needed to talk to. The CliffsNotes version of my conversation with him would be something like this: It's all about the waves of motion traveling from your hand, through the rod, down the string, and to the lure at the end. (Also, from my rudimentary understanding of physics, I believe that a cat in motion will stay in motion until they are tired, at which point they lie down and take a nap.)

Dr. Gbur pointed out that when you move a longer rod, the movement will be magnified at the toy end. Because the weight of the rod is negligible, waves can travel down it easily. And the longer the rod, the less effort you have to put in to increase the movement of the string and toy at the other end. Is a long toy for the lazy?

A more flexible rod will allow for more variety and perhaps more interesting movements (think wounded bird or confused mouse). The waves of motion will be delayed due to the flexibility of the medium, resulting in less predictable behavior on the lure end of things. This could be good or bad depending on your intentions and what your cat enjoys.

Now that we know a little physics, we can correctly conclude that a shorter, or more rigid, rod allows for more precise, controlled movements of the toy. This is great for those moments when you are barely moving the toy, while your cat is completely focused on stalking.

The strings or ribbons connecting the rod to the lure shouldn't be too long or slack; they need tension to propel the movement from your hand into the lure. For example, a heavy lure and a light, long string would be problematic. The toy would just flop around sadly or barely move at all. To control the movement, you need more momentum in the string.

And while we're talking physics, let's have a word about laser pointers. If you choose to use them, keep the laser light's movements in the realm of possibility whenever possible. Failure to do so is just one reason that laser pointers are a less than ideal play object. The cat can never catch their prey, and people often move the laser in an unnatural way—for example, moving the light halfway up a wall and then abruptly turning the laser pointer off. The prey just vanishes into thin air unexpectedly. If you must lase, then keep it limited, keep the movements realistic, and switch to a physical toy at some point during the session so your cat can have the satisfaction of physically catching a toy. Although dogs are prone to laser light obsession, to date this has not been identified as a major problem for cats; however, I recommend laser pointers only as a supplement for tough-to-warm-up kitties, and certainly not as a main element of your toy tool kit.

A recent study also showed that cats understand some basic elements of the laws of physics: perhaps further support for the premise that cats are plotting to take over the world. Thankfully, we don't need a PhD in physics to use cat toys effectively. But it definitely doesn't hurt to have a cat-loving friend with a PhD in physics, like Dr. Gbur, to help you understand what you're doing.

DIY AND NONCONVENTIONAL TOYS

How many times have I heard cat owners complain about all the money they spent on toys that their cat won't play with? Well, let's look at some options that are essentially free: don't overlook household (or outdoor) items that might serve as excellent lures for your favorite wand.

You've probably caught your cat playing with things that you never would have considered a cat toy—juice cap rings, pencils, bottle caps, tea bag tags, empty spools, wine corks, toilet paper rolls, corn husks, leaves from your yard, and baby shoes all come to mind! Some of these things can easily be turned into an interactive adventure by attaching them securely to a wand with a little string or even a plastic twist tie (just make sure your cat won't chew on the twist tie). I especially like to attach small pieces of paper (kraft, tissue, newspaper, what have you), leather, felt, or denim to a swivel to create a homemade temporary toy.

If you're super crafty (unlike me), you can even build your own entire cat toy—from rod to string to lure. If you look up "DIY cat wand toy" in your favorite search engine, you will find lots of videos and instructions on making toys. I've seen cool toys made with mesh and latex tubing, and the sky is the limit if you have a glue gun.

Don't rule out what might be sitting in your pantry. My cat Coriander occasionally enjoys playing with potatoes, and a green bean is enough to spark excitement among all three of my cats. They will bat at it, fetch it, and even chase it when it's attached to a wand toy. Vegetables smell organic, plus they are small and prey-like (green beans are even a little wormlike), and the nice thing is that you can compost them when your cat is done with them. (Note: When looking for vegetables that can double as toys, avoid using garlic, onions, chives, scallions, or leeks as they are toxic to cats.)

SKIP THE LINE AND GO STRAIGHT TO THE LURE

For the ultimate control over your lure, try attaching it directly to the end of the toy wand (no strings attached) with a jewelry clasp or wire-free plastic twist tie. This works well if you've got the type of wand that has an eye hook at the end (you can also make one by screwing an eye hook into the end of a dowel). Many lures have a small loop at one end that allows you to connect it to the wand. By skipping the string, you'll find a unique type of movement—more subtle than a toy at the end of a string but also with much more control. An excellent toy modification for cats who especially love stalking.

SAFETY FIRST

When you are not directly using an interactive toy with your cat (in other words, when your hand is not attached to it), *it should be put away where your cat cannot access it.* Cats' tongues are structured with pointy barbs that face the back of their throats, feeding anything they swallow right down the gullet. Despite the pervasiveness of the ball of yarn in cat culture, it is not a good cat toy. Small objects and strings can present a choking hazard or lead to a deadly obstruction in the intestinal tract.

Watch for toys with bits that can be chewed or ripped off, such as glued-on noses or eyes. Use common sense with household objects. There are some "dangerous" items that may be okay under close supervision, such as a shoelace. If you are always holding the shoelace and you put it away when your cat isn't playing with it, you're all good. But in general, avoid leaving your cat unsupervised with any very small items, *especially ones they seem incredibly attracted to like hair ties and cotton swabs.*

Toys can also present a strangulation risk: a cat can get tangled in a string or wire. If that string or wire gets caught in something (say, underneath the leg of the couch), it can literally strangle a cat if it's wrapped around their neck and they struggle to get away and pull on the wire (unfortunately, I know someone whose cat this happened to). So please, please come up with a system for storing your toys that your cat cannot break into. I have a few suggestions:

- You can lay interactive toys flat in a dresser drawer, storage bin, or an under-the-bed sweater bin.
- A cylindrical umbrella holder or a freestanding toilet paper roll holder can store several interactive toys upright (place the holder in a closet).
- Try a fishing rod rack or, as much as I hate to say it, one of the cheaper gun racks.

TOY ORGANIZATION

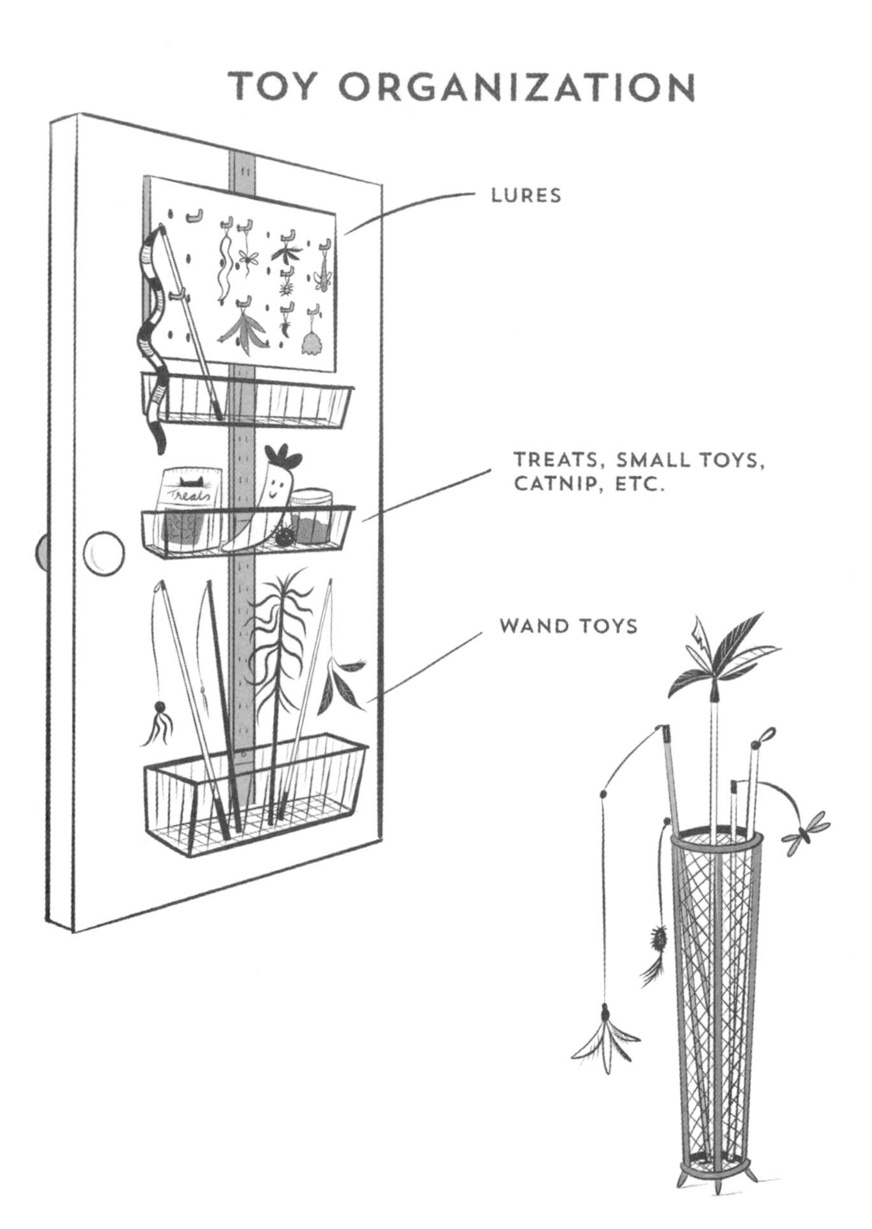

- Stores dedicated to organization and storage often sell nifty modular units that you can customize to your toy needs. Mounted closet organizers offer several options such as baskets for storing extra lures, and vertical racks for storing the rods and longer toys.

Keeping your interactive toys organized and keeping your cat safe prevents a trip to the emergency room or a very expensive surgery. Remember: Your cat's brain is the size of a shelled walnut and they don't always know what is best for them. That's what we are for.

CHAPTER 5

Setting the Scene

Just as theatrical productions have props and actors, they also have a stage and set. In the theater, the set is composed of movable and stationary items that set the scene and create illusions. When it comes to cat play, the toys are your props, and you and your cat are the actors. Now let's set the scene for your cat.

Your cat is a wild hunter; in their mind, they might be crouching in a savanna, like the homeland of their ancestors. They might be climbing a tree to chase a bird. They might be hiding behind a shrub, stalking a butterfly. They might even be starring in a sci-fi film that allows them to defeat the laws of gravity, climb walls, or slay dragons. Your job is to help them live out the movie playing in their head.

Cat toys are the starting point, but we can do more. Look around you; your home is filled with dozens of available objects that can supplement your play session. Even your recycling bin has something to contribute! Although this guide is non-exhaustive, I hope it's a starting point for you to recognize how you can expand your play world by thinking about your cat's environment just as much as you think about the toys.

MAKING SPACE

I did a consultation once for a man who had a very young, playful kitten and a very small studio apartment. The tininess of his apartment was compounded by the fact that he had a terrible hoarding problem. There were literally piles of books and paper and boxes covering almost every square inch of the floor, bed, and tables. Any surprise his cat was keeping him up all night and biting him at every opportunity? There was really nothing else she could do.

I like to have a relatively large open space for play that allows for backflips and stellar leaps, not to mention space to set up boxes, tunnels, and other temporary play structures. I would say the minimum play space is approximately six feet by six feet, but ideally you have an open space that is even bigger, at least during your play sessions. Feel free to move coffee tables, couches, chairs, and the like out of the way to clear a little area for your cat.

Even if you don't have one large play space for your cat, think about how hallways and other spaces can connect different parts of a play session together. Perhaps you start playing in the bathtub, then spend some time in the hallway, eventually leading your cat to the living room, with a grand finale on their cat condo. You've created more space by creating a maze of sorts that you and your cat move through.

You can also gain square footage by thinking in three dimensions. Incorporate both cat and human furniture into play to make it more aerobic and dynamic. You will have to determine your "house rules" for where your cat is allowed and whether you want your cats to play in certain areas. For example, in our house, the couch and dining table are fair game, but the cats are not encouraged to play on the kitchen counters, mostly for safety reasons.

Think carefully about whether you will let your cat play on your bed. Now that my cats are adults and sleeping through the night, I do allow them to play on the bed (it's soft and great for jumping, and for hiding toys under blankets . . . and my cat Professor Scribbles really loves to play there). But it is possible that for some cats, you may be sending a message that it's okay to play on the bed, and your cat may not discriminate whether humans might be sleeping in it when they want to play (and if your cats are keeping you up all night, keep reading, as I'll be discussing how to get some decent sleep in chapter 11). There's no one right answer, just what's right for you, your cat, and your precious sleep.

Depending on your cat's style of play or what activities you shall partake in, you may want to offer both carpeted and slippery surfaces for play. Consider how the toy you are using may move across the floor—will it slide and glide or will it catch and drag? (Both can be viable options!) Then think about how your cat may move across the floor. For example, carpet (or an area

PLAY SETUPS

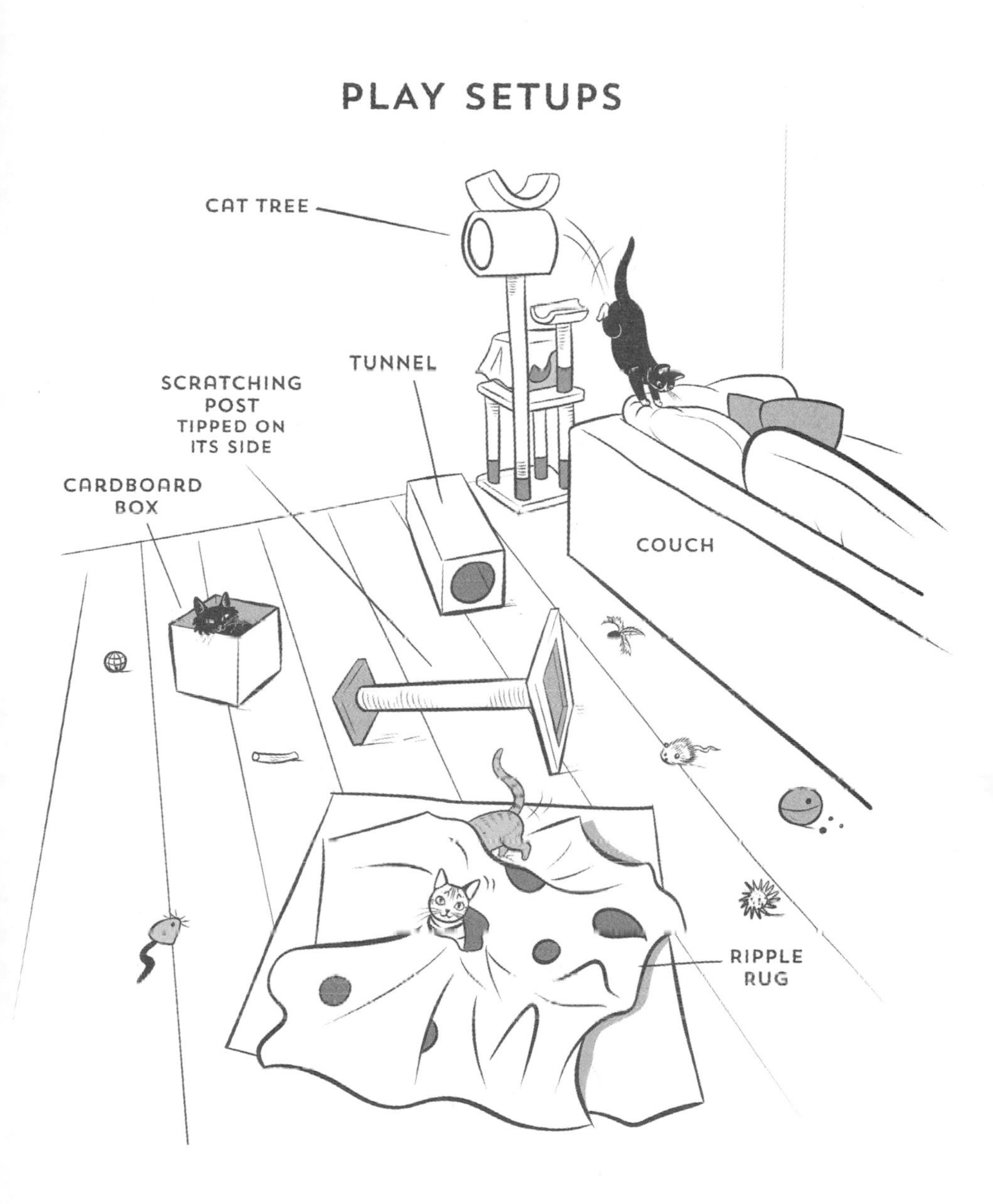
CAT TREE
TUNNEL
SCRATCHING POST TIPPED ON ITS SIDE
CARDBOARD BOX
COUCH
RIPPLE RUG

rug) is great for stability when cats are doing a lot of leaping and backflips. I've known a few cats who like to drag themselves around by their claws a little bit when they are playing as well.

A hardwood floor can be great for on-the-back play or when using cat beds for part of the play. But for older cats, or cats with any joint or hip issues, a slippery floor may not give them enough traction to feel comfortable. If that seems to be the case, play in a different area, or use an area rug or a play mat to help your cat feel more comfortable.

THE PROPS

Let's start with the stuff you already have: human furniture. You can encourage your cat to explore three-dimensional hunting by enticing them to track a toy up onto your couch or a chair. Depending on the rules in your house, tables may also be included. But if your cats aren't allowed on tables, don't despair. Those tables have legs, you just have to know how to use them! Trail a small toy slowly on the ground until it is partially or completely hidden behind a table leg. You can even tap or scratch the toy against the surface of the leg to make a small rustling sound that your cat will find alluring.

YOUR TURN!

Move or use some furniture to create a new play space for your cat. For example, set up a row of chairs that allows your cat to leap from one to another (adjust the distance between them to accommodate your cat's jumping abilities) like indoor parkour.

Many cats already use the backs or ledges of couches as a freeway to get from point A to point B. Now you can make it an official part of your play pathway. Couches also have cushions, which are perfect for hiding toys behind or between. One seventeen-year-old cat I worked with had lost her interest in play—until we moved the toy very slowly between some of those superfluous throw pillows that don't serve much function, except apparently to get geriatric cats inspired to move their bodies. And there's no reason not to build a couch-cushion or pillow fort for your cat. It provides a perfect hunting blind for your cat to ambush from.

HUNTING BLINDS OR HIDES

A hunting blind is a structure that hunters use to prevent themselves from being detected by prey. Humans who hunt often use or build structures they can hang out in, perhaps even for hours, waiting for a deer or duck to pass by.

For cats, a hide is anything that provides them with cover while stalking. Outdoors that could be tall grass, a bush, or a tree, or human-made structures. When playing with your cat, think of places in your home that could provide your cat with partial or total cover and serve as a hunting blind.

YOUR TURN!

Create a hunting blind for your cat: something they can hide in, under, or behind while stalking. Some options include a towel draped over the sides of a kitchen chair, a cardboard box tipped on its side, or a potted plant, moved to a new location, which your cat can use for cover.

Hopefully your home also contains some cat furniture. If you have cat trees or perches, you can include them in your play session. Vertical space can increase cardiovascular activity during play to tire out those tough-to-exhaust hyperactive cats. Get your cat to jump up to follow the toy, jump down to follow the toy, repeat again and again. Pet steps and ramps can provide additional excitement and pathways to different vertical spaces, and they can be especially helpful for older or differently abled cats who might need a more gradual or safer way to access elevated spots (more on this in chapter 12).

I highly recommend tipping a scratching post onto its side for play. A lot of cats will intuitively lie on their backs and grip the post with their front legs, kicking it with their back legs. You can move a toy along the post to get your cat started, but they may just enjoy the sensation of kicking, minimal toy interaction necessary. In our house we call this *pole dancing* because it often involves lots of spinning, twirling, and leg extensions.

There are several fantastic cat tunnels out there, and pretty much any of them can enhance your cat's playtime. They can be a simple tube with an opening at either end, or a more elaborate X- or Y-shaped structure that allows multiple entry and exit points. Some have several peepholes in them, whereas others only have the main entries. Crinkle tunnels are filled with a fabric that causes a rustling sound when touched, which excites many cats. Some pet companies also make lightweight nylon pop-up tents, play cubes, or huts. These tents are usually opaque with multiple openings, allowing for stalking and pouncing, and they fold up neatly when you are finished playing.

Don't forget all those cat beds sitting around that you complain your cat never sleeps in. If nothing else, a cave-style bed or one with high sides can serve as a hide from which your cat can stalk toys as you move them. Your cat might also like playing *in* the bed.

Try a cave-style bed with two openings—one large and one small. I might be imagining it, but I swear my cat Coriander likes

to stare through both holes, to focus on a moving toy at the far end—much like an aperture focuses an image. At the perfect moment, she will dive into the cave and out through the other end to grab the toy.

A bed with a soft and smooth bottom is a great choice for dive-bombing into and sliding—with enough momentum and a clean hardwood floor, it can glide several feet.

REUSE BEFORE YOU RECYCLE

Sometimes the best cat toys in life are free: I don't think I need to tell you that cats love cardboard boxes. Boxes of all sizes and shapes can add to your cat's indoor jungle gym, and when your cat loses interest, or the boxes start to fall apart, you can just toss them in the recycling bin.

For many cats, just setting an empty box in the middle of a room is enough to get them to investigate, jump inside, and wait for something to happen. I've seen cats squeeze into boxes that were much too small, and I've seen them enjoy boxes that were much too big. You can put a smaller box inside a larger box and blow their tiny feline minds. Really, have fun with the boxes. My favorite thing to do is fill a cardboard box with tissue paper or kraft paper (the stuff that all of your cat food orders seem to be packed in); cats love the crinkling sound and may even "nest" inside the box—after they've enjoyed a thorough play session in there first. You can also toss a few treats in the box and allow your cat to dig through the paper to find them.

YOUR TURN!

Save up at least three medium- to large-sized boxes to build a play structure for your cat: cut some entryways in them and use packing tape to secure them together into a fort. Or use them to create tunnels and pathways; add some interactive playtime and most cats will not be able to resist!

And tissue and kraft paper do not need a box to provide fun for your cat. Yes, I know it's going to sound like my house is a mess, but we often have tissue paper and large pieces of crumpled kraft paper strewn across the floor forming makeshift tunnels. Because kraft paper is heavier and holds its shape, these quasi tunnels are perfect for hiding in, or for providing passage from one cardboard box to another. Tissue paper is more delicate but is good for moving toys underneath, creating an illusion much like a mouse under a rug. Some cats also enjoy ripping and shredding tissue.

Before we figured out that cats loved cardboard boxes, we thought they loved playing in paper bags (remember those?). Although we may not have as many paper bags in our homes these days, they still make a fun prop for play—just cut off the handles so that your cat does not get the bag stuck on their neck. As always, safety first!

THE SOFT STUFF

Mice love carpets, and by proxy, so do cats. Perhaps Tom and Jerry got us all thinking that cats chase mice who are running under rugs. The reality is probably a little different, but according to several pest control websites, mice do in fact love rugs. Mice apparently chew carpets to tear away fabric fibers for building a nest—and a rolled-up carpet is a perfect spot for hiding.

TOY UNDER THE MAT

We can create a "mouse under the rug" experience for our cats by moving the end of a toy under a towel, blanket, or even a rug. Cats especially seem to love when a toy moves just under the edge of a rug or towel, eluding capture yet still detectable through the fabric. If you want to go all out, I highly recommend getting the SnugglyCat Ripple Rug, which is a two-layer rug that can be reconfigured into multiple shapes. The bottom, solid layer has an anti-slip rubber backing to keep it secure. The top layer has multiple openings and Velcro bits that can be attached to the bottom rug for reshaping it into a three-dimensional play zone with tunnels, holes, and hills for hiding and playing in.

IN THE MOOD

The feline equivalent of putting on romantic music in a room lit by candles might be putting on a bird video or even just the sound of birds chirping. We can also experiment with playing with our cat in the semidarkness! A great way to add a little variety to your play sessions is to turn down the lights. You can also try playing with your cat at dawn or dusk to mimic the times they would naturally be interested in hunting.

DON'T RULE OUT . . .

Have you tried playing with your cat in the bathtub? It's like the ultimate cardboard box, just made of porcelain and with fantastic curtains for stalking behind. A tub is also great for rolling a few Ping-Pong balls in, or for cats who enjoy playing with water dripping from a faucet.

And although many cats have learned to hate their cat carrier, there's no better way to start fresh and help them learn to like the carrier than by leaving it out all the time, plunking some comfy bedding and treats in there, and using a toy to get them to play around (and perhaps eventually *in*) it.

Finally, if you are lucky enough to have a catio or similar outdoor enclosure, the fresh air, sounds of birds chirping, and the feeling of the breeze are sure to excite your cat's inner hunter in a safe but semi-outdoor play adventure.

New places and props for playing with your cat will help you get your "off-off-off-off Broadway" production (and your cat) off the ground. From your cat's perspective, all the world is a stage, and we are merely players, in every possible sense of the word.

CHAPTER 6

Playing for Good Health

Now that I've addressed how, where, and when to play with your cat, let's get into *why*. I'm arguing that, now more than ever, cats need playtime. Why? Because many cats are overweight, bored, and frustrated, leading to health and behavior problems, which put them at risk of losing their homes or even their lives. Many cats are also silent sufferers of a restrictive lifestyle. What starts off as boredom and frustration can turn to apathy and learned helplessness. Let's look at where things have gone wrong for our feline companions and where interactive playtime can right the kitty ship.

THE FELINE OBESITY CRISIS

Almost 60 percent of pet cats are overweight (meaning they are at least 10 to 20 percent over their ideal body weight). Although at some point humans decided that fat cats are cute, being overweight is the most common *preventable* health concern in owned cats. Obesity leads to several other health problems, such as an increased risk of diseases like diabetes mellitus, cancer, and heart disease; orthopedic problems like hip dysplasia and arthritis; and urinary tract problems. Ultimately, being overweight or obese shortens a cat's life span and greatly reduces their quality of life, since they feel crappy, are in pain—oh, and they can't reach their asses to clean them anymore. Not sounding so cute now, huh?

How did cats get so overweight in the first place? First of all, spaying and neutering our pets has effects beyond just preventing the birth of puppies and kittens. Removing the gonads decreases the metabolic rate, meaning that neutered cats need fewer calories to maintain the same body weight. At the same time, neutered cats want to eat more food. And once the pounds are on, even if they're fed fewer calories, altered cats will continue to carry more body fat and weight than their gonadally intact equivalents. Voluntary activity levels also tend to decrease after a cat is neutered. The combination of slower metabolism and less activity makes it easier for our cats to pack on some extra weight, and once cats are overweight, they become even less active relative to leaner cats.

It's generally agreed upon that neutering or spaying our cats is best because it prevents the birth of unwanted kittens, and the procedure reduces some undesirable behaviors associated with mating, such as urine marking and excessive vocalizing. But essentially, it makes our cats lazy, and this may be further compounded by keeping cats exclusively indoors. Many studies have shown that cats with restricted or no outdoor time are at greater risk for being overweight or obese. However, much like neutering and spaying are for a greater good, the consensus, at least in the United States, is that keeping cats indoors only is safest for them, and also for the animals they prey upon. The downside is that the indoor-only lifestyle may further restrict activity, when our cats are already on their way to full-on couch-potato status. Interestingly, we have very little data to suggest that outdoor cats are out there running marathons. In other parts of the world, such as the UK, cats are much less likely to be kept indoors only, but the percentage of the overweight cat population in the UK is similar to that of the United States. So even if your cat goes outside, there's no guarantee that they're doing anything much more strenuous than their indoor counterparts.

Being "free-fed" (meaning a bowl of food is always available) is one of the biggest culprits for this feline health crisis, yet although most obese cats are free-fed, not all free-fed cats are obese. The relationship between feline fitness and food is complex.

The irony is that what seems to get cats active is mealtime! Food anticipatory activity (FAA) has been measured by comparing how active cats are in the two hours before their scheduled feeding time with the rest of their calendar. Multiple studies show an uptick in activity during this time, which is more intense when the number of meals per day is increased.

You may have observed some FAA in your own cats if they are fed on a schedule. The rubbing, the pacing, the rolling, the flirting—who knew that being hungry makes cats more active (and apparently also more loving)? If your schedule permits, there's no harm in feeding your cat more frequent, smaller meals (and try playing with them before you throw down the food, to take advantage of that FAA!).

The activity patterns of cats are also influenced by human behavior; one study found that laboratory cats were more active on the weekdays, when they received more human interaction, compared to weekends, when they received a more condensed care regimen. So don't underestimate your ability to help your cat burn a few calories just by being around and by interacting with them.

If your cat is already overweight, just playing probably won't be enough to get them back into shape. The best thing you can do is work with your veterinarian to come up with an appropriate feeding plan and ask if your cat is healthy enough for interactive play.

As you can see, the odds are a bit stacked against our cats staying fit when we're just trying to be responsible pet parents by getting them fixed and keeping them indoors. But rather than throw our hands up—we should tackle this problem head on! You can use more frequent feeding of small meals and interactive play, timing both to optimize the effects, to help your cat lead a healthier, more active lifestyle.

STRESS, BEHAVIOR, AND WELFARE

In addition to the feline obesity epidemic, a lot of people think their cats are naughty. When cat owners are asked innocent questions about their cat's behavior, over 60 percent will state that their cat has one or more undesirable habits. This isn't just about minor nuisances like knocking things off your shelves—we're talking peeing on the bed, biting their humans, fighting with other animals in the home, and keeping their owners awake all night. Things that many humans find annoying and hard to live with—the kinds of behavior problems that send cats to the animal shelter.

Behavior experts agree that significant causes of companion animal behavior problems include boredom, frustration, and stress. In fact, many behaviors that we see as problematic are just normal responses to an inadequate environment. Think about it—although we have domesticated cats, we have not really changed many of the behaviors and instincts they inherited from their wild ancestors. But we brought them inside, plunked a bowl of food on the floor, and asked them to "please, please, oh please be a low-maintenance pet." We forgot about the fact that they naturally should be hunting fifteen to thirty times a day and instead expected them to be content with chasing a dust bunny across the floor once in a while. As a consequence, many cats experience chronic stress, which is associated with depression, behavior problems, and health problems.

Negative emotional states have also been associated with

overeating in several species, and emotional overeating has been proposed as a coping mechanism that can alleviate feelings of stress in our pets. Social isolation, depression, and deprivation of play all increased food consumption in laboratory rats. Obesity may even be a warning sign that cats are experiencing negative emotional states. We can see a vicious cycle developing from a complicated relationship between stress, obesity, and behavior problems.

Can play help cats overcome those negative emotions? Many humans exercise for the endorphin rush—people who exercise regularly experience fewer depressive symptoms and many report a sense of euphoria after a run or other intense physical activity. Research in animals suggests that play can lead to similar pleasurable states, reduced feelings of anxiety, and may even prevent health-related issues in our cats. A 2014 survey of cat owners also found that cats who received more daily playtime (at least five minutes per day) had fewer reported behavior problems than cats who only received one minute of playtime each day. Dusting off those interactive toys or buying (or making) some new ones is an important step in helping our cats stay mentally healthy and well-behaved (and I'll talk specifically about using play to address different behavior issues later in the book).

Play behavior is often used to assess whether an individual is experiencing positive welfare—if you see it, a cat is likely living in an environment that feels safe and meets their basic needs. We can also learn about the importance of play by observing when we tend to *not* see it—and as research has found, there is decreased play in cats who have been declawed, who are housed in animal shelters, or who are in situations where there are more cats than a space can safely support. Although play cannot be the only metric we use to measure whether a cat is having a good life, it is still a vital one. Perhaps most importantly, there's nothing that would suggest that playing with your cat can decrease their quality of life, especially when done with care and tailored to that cat's needs.

SIGNS OF GOOD WELFARE IN OUR CATS

People often ask, "How can I know if my cat is happy?" Dr. Andrew Fraser, in his book *Feline Behaviour and Welfare*, proposed this list of "welfare indicators" (in order of significance! Note what is *number one* on the list). Place a checkmark next to each indicator your cat shows regularly:

- ❑ *Daily play, chasing, racing, and climbing*
- ❑ *Stretching*
- ❑ *Yawning*
- ❑ *Self-grooming*
- ❑ *Rubbing head and body against the owner*
- ❑ *Responsiveness to petting*
- ❑ *Purring when petted*
- ❑ *Vocalizing gently when greeting the owner*
- ❑ *"Getting underfoot"*
- ❑ *Elevating the butt or tail in response to touch*
- ❑ *Using elevated surfaces or perches*
- ❑ *Licking other cats or the owner*

SIGNS OF GOOD WELFARE

DAILY PLAY

STRETCHING

YAWNING

SELF-GROOMING

RUBBING HEAD AND BODY AGAINST THE OWNER

RESPONSIVENESS TO PETTING

PURRING WHEN PETTED

VOCALIZING GENTLY WHEN GREETING THE OWNER

"GETTING UNDERFOOT"

ELEVATING BUTT OR TAIL IN RESPONSE TO TOUCH

USING ELEVATED SPACES OR PERCHES

LICKING OTHER CATS OR THE OWNER

BUILDING A BOND

One of the most important reasons to establish a play routine with your cat is the positive impact it can have on your relationship. Cats are very much creatures of habit and like things to happen in a predictable manner. By establishing regular playtimes, you are helping your cat have a sense of stability in their day, in addition to the health benefits we've already discussed. Play is fun, and they will associate that fun with you, and even look forward to your sessions.

Sometimes play is the best way to bond with your cat. Not all cats are cuddly or enjoy a lot of touch. That can leave humans at a loss; instinctively, many of us want to pet our cats. They are soft and cute. If we can't pet them, well . . . why do we even have them? The answer to that is complicated, and it's often referred to as "unconditional love." Unfortunately, we are kind of geared to frame our relationships with our pets around "What's in it for me?" instead of around "How can I make this little creature I chose to bring into my life happy?"

If your cat is one of those who is not particularly thrilled with being held or with a lot of petting, it's time to think outside the box. Sometimes even just backing off with the constant handling is enough to help them feel a little more relaxed around you. Positive reinforcement training, a method that rewards cats for good behaviors, is another great way to interact and build trust. But I find interactive play to be an especially powerful way to build a bond with a cat. You can keep your distance if that is what your cat prefers, or use the toy to get them comfortable being closer to you.

And hopefully interactive play will help you appreciate some different aspects of your cat's personality and self—perhaps some things you may have overlooked when you dismissed them as aloof or too independent. You will see their eyes light up when they focus on the feather wand as if it were a real bird; you will see how every muscle in their body tenses in anticipation of an attack. You will see your cat in a new light: With play, you may see your cat as brave and powerful, or athletic, or even goofy and uncoordinated. But when done properly, a play session should leave you laughing, or at least with a smile on your face, and should leave your cat feeling satisfied and at ease.

CHAPTER 7

From Kitten to Seasoned Adult

HOW PLAY DEVELOPS IN CATS

Play is for life, but knowing the feline life stages can help you adjust both your expectations and your play technique so you can meet your cat's needs no matter their age or abilities. You've probably heard that each cat year is something akin to seven human years. Unfortunately, the math isn't that simple, but rather than trying to calculate whether your cat should be able to legally vote or drink, or is technically older than you, let's just take a look at the different feline life stages, on cat time.

KITTENS: BECAUSE THERE'S NOTHING CUTER

Kittens are considered as such from birth until they are a year old. Because there's possibly nothing cuter than studying kittens, the 1970s and 1980s brought us a huge body of research investigating the play behavior of kittens—and probably led a lot of people to thinking that getting a PhD was a good idea. (Is playing with kittens what scientists really do? Yes—some of them!)

From this body of work, we have some clarity about how play behavior develops in kittens, especially in very young kittens, as most of the developmental research has focused on the first few months of a kitten's life. But it's also important to recognize that some studies were on very small numbers of kittens, so while we can use this research as a guide for our understanding of play behavior in cats, you shouldn't be concerned if your kittens don't quite follow the rule book to the letter. When we do animal behavior science, we are trying to generalize about a population (all kittens) from a smaller group (the kittens we studied), and so the aggregated results don't necessarily represent what happens for every single kitten.

Kittens are born pretty much helpless: blind puffs with bad hearing who can crawl around and find Mom by smell. But by two to three weeks of age, kittens' eyes and ears are open, and they are starting to move around with some skill. This is when kittens begin playing, and for the next four to six months, play blocks off a lot of time on a kitten's calendar. In fact, a 1984 study parsed out just what kittens do all day and found that 9 percent of their time was spent in play. Now perhaps that sounds like a pittance, until you realize that translates to about two hours a day! (Note: I cannot tell you with any accuracy how much time your kitten will spend staring into space [we call this "the empty thought bubble" in our house] or what amount of time they will watch you try to take pictures of them [often unsuccessfully] for your Instagram account. But we do know they will also spend a lot of time sleeping! Play hard, sleep hard!)

TYPES OF PLAY

Play is often categorized by what the kitten's rambunctious behavior is directed toward, such as *social play*—which typically occurs toward whoever is available to play with, usually other kittens or cats; and *object play*, which is directed toward toys or other household items. *Predatory play* walks that fine line between hunting and uncertainty, where kittens are learning to kill. Finally, there's also *locomotor play*, where kittens explore the 3D world around them by climbing, leaping, and, sometimes, by falling down.

FOUR TYPES OF PLAY

FIRST THINGS FIRST: PLAY WITH OTHER KITTENS

Of these types of play, social play is the first to emerge, right when kittens are showing enough coordination to start walking. Social play might arise first because if you are a kitten, your littermates are readily available—you are all wiggling around in the same nest together. Single kittens, lacking littermates, try to play with Mom instead, much to her irritation (on average, mothers spend less time with a singleton kitten than they spend with kittens of a larger litter. Moms just don't want to play with their kittens, making them a poor substitute for littermates in that regard).

Some aspects of social play look like behaviors kittens will later direct toward toys or prey—with lots of stalking, chasing, and pouncing, just like they might do with a mouse. Other play behaviors, such as swatting, wrestling, biting, and raking with the back legs, resemble a "gentler" form of fighting, perhaps preparing kittens for the potential need to defend their turf or negotiate access to resources down the line. The development of behavior makes no predictions about the future; the underlying instincts are present whether that individual kitten will have to hunt to survive, or whether they will be friendly or ferocious with other cats. Instincts just give kittens the practice they need to be prepared for the worst-case scenario.

Social play also has some unique features—signals that a kitten can use to say, "I'm playing with you!" Side-stepping, a belly-up pose, and a mouth held slightly open are some of the most common ways that kittens announce their intentions—and other kittens frequently respond by mirroring these behaviors. Signals such as these allow kittens to let others know they are ready to play, increasing the chance that another kitten will take them up on the offer to romp.

This romping between kittens continues with vigor until kittens are approximately four months of age, at which point it begins to wane. Research by Dr. Meredith West on the social play of kittens found that beyond this age, kittens spent less time playing and more time sleeping or in a "quiet, alert" state.

As kittens get older, social play may even become aggressive or unfriendly, especially when cats are approaching sexual maturity; most studies of play have been of kittens who were not spayed or neutered. The average pet cat is spayed or neutered, often by two to three months of age, which changes many aspects of their behavior, such as reducing their desire to protect territory and fight. As cat owners and adopters, we often separate kittens from littermates at this age, as families are frequently broken up when kittens are adopted into different homes. Instead of sticking with siblings, they often must integrate with unfamiliar cats or adapt to life as an only cat.

THE OBJECT OF MY AFFECTION: PLAYING WITH SMALL THINGS

Object play starts more slowly and a little bit later, ramping up closer to the time that kittens are weaning (around seven weeks of age), and peaking when kittens are approximately four to five months old. This is also the time when play patterns begin to change; it turns out that playfulness in kittens from four to seven weeks of age (when play is very social) is not a good predictor of how playful they will be from eight to twelve weeks of age (when they primarily want to play with things). This strengthens support for the idea that social play and object play are distinct behaviors with different motivations, and not necessarily related to each other (in other words, how much romping you did with your littermates does not predict how much you love batting Ping-Pong balls around!).

When kittens engage in object play, there is a bit of exploration (sniffing, cautious touching), play (grasping, tossing, and biting), and some behaviors that are in between (running at the object but also stopping to explore it before touching). In one study, kittens who were presented with new objects (such as wine corks, Ping-Pong balls, and dead mice) spent most of their time in exploration mode, and only played vigorously with items once they were more familiar with them. Much in the way that demisexuals feel attracted to someone only after they've gotten to know them, perhaps we should call kittens *demipredatory.*

Once they are fully weaned, kittens are often left by Mom to be self-sufficient; her training and the time they spent playing have hopefully served their functions in preparing a kitten for survival. But now, object play shifts to predation of real mice and birds. There's less time to play with Ping-Pong balls when you're a four-month-old who must focus on hunting to survive. Kittens living a pampered, indoor lifestyle with a full bowl of food likely continue directing that hunting energy toward toys, especially when lacking any live prey to attack.

OBJECT-IFICATION: PREDATORY PLAY

When play behavior is directed toward prey (whether living or dead) instead of a toy (aka object), it's called *predatory play.* In kittens, the behavior patterns look much like object play. Prey animals can be threatening to kittens, just like they are to adults.

When kittens are interacting with live prey, they may have difficulties killing them. Sometimes this difficulty gets labeled "play," although it's far from clear that the kittens are actually playing and not just struggling. They might touch or poke the prey animal, walk away, roll around with it, or even toss the prey animal in the air. When cats are hunting, what looks playful can sometimes just be self-preservation.

LOCOMOTOR PLAY

Kids love to gallop and skip and wiggle and dance for no clear reason. Wandering aimlessly—I think we've all done it, whether in a forest or in a shopping mall. Kittens do it too—sort of. When animals (mostly young ones) move around without purpose (meaning not moving to interact with objects or other individuals, and not trying to explore *or* escape), it's called *locomotor play*.

To assess the development of locomotor play, researchers provided kittens with a climbing structure that they could interact with daily for thirty minutes at a time. Kittens were thirty-six days of age when they started, and at first, they stuck to the lowest levels of the structure. Once they were forty-eight days of age, they immediately climbed on the jungle gym, attempting to climb higher as they got older. Although some kittens lost their balance, few of them completely fell off the jungle gym, and those who did dusted themselves off and got right back up again, seeming only slightly dazed. (Note: if you provide climbing structures for young kittens, consider doing so in carpeted areas!)

The takeaway: Since there was no immediate reward (such as

food) associated with climbing, it appeared that climbing in and of itself was fun for kittens. We know that cats love to be up high! Given that Canopy Cat Rescue, certified tree climbers in Washington State, rescues over five hundred cats a year from trees, we can see that what goes up doesn't always easily come back down. However, consider the benefits of providing some safe climbing options within your home, such as cat trees and cat-friendly shelving.

THIS IS THE REST OF YOUR LIFE

And that's it when it comes to our understanding of the development of play in cats. The playfulness of older kittens is sadly not well studied, and even less is known about adult cats; when looking over the scientific literature, it's almost as if kittens just fall down and stop playing at precisely six months of age. Anyone who lives with cats will tell you this just isn't true. What has led to this significant research gap?

Perhaps a scientific bias has prevented further study of play in adult cats; if we wrongly assume that the only function of play is to prepare an individual for adulthood, then play in adult animals gets relegated to the "behavioral fat" category (I kid you not, this is how scientists have described it). Play has been treated as just child's play rather than viewed as something worthy of lifelong study or that has value to adult animals.

Cat owners may also be contributing to the problem, because cats aren't given sufficient opportunities to play. It's difficult to study what we don't see. And although many cats live with other cats, approximately a third of them grow up as solo pets, meaning that opportunities for same-species social play become nonexistent for this segment of the feline population.

LIFE STAGES: THROUGH THE YEARS

There's no real rite of passage when a kitten becomes an adult. At one year of age, kitten food disappears, adult food appears, and now cats are considered grown. In nature, they would be independent and fully reproductive for *several* months by now, but I believe that the mind of the cat stays kitten-like for a few years at least, the period when cats are technically considered "juniors."

As the kitten becomes an adult, less time is spent playing as they shift their focus to hunting. Adult cats become quite efficient at hunting but spend more time each day roaming in search of prey. Their overall amount of daily activity stays close to the amount of time that kittens spend in play—again, up to two hours a day. Of course, their total amount of active time is usually spread out over the course of several hours, in several short bursts of energy, with frequent naps or rests in between (how do I get this job?).

When cats are around seven years of age, they are now "mature," or some might even say middle-aged. There's more of the feline equivalent of going to bed early after a night of television on the couch. Your cat will still be interested in play, but it may be of a much less athletic type (expect fewer backflips and more "playing while laying"), and your cat may be satisfied or even done with shorter play sessions. Cats with declining interest in play may benefit from some shifts in their daily routine that still keep them active and engaged, such as trick training, harness walks, or increased sensory enrichment. They may enjoy more snuggles (we can all dream!) or appreciate a heated pet bed more than ever. But the play doesn't disappear, it just shifts a little.

At ten, your cat is officially a senior, and at fifteen they are considered geriatric, or a "super senior." Although cats are living longer than ever, an increased life span can bring challenges. As they age, cats become more susceptible to cognitive dysfunction, a condition like Alzheimer's disease in humans. Senior cats are also more likely to have chronic diseases that may slow them down. That said, just like humans, cats age individually. Some cats become frail at a younger age and some cats are robust until the very end. There's no reason that a super senior can't be a super player!

Although older cats may have physical and cognitive challenges, play is just as important as ever. Even in short amounts, play can serve as both a preventive and a treatment for health and cognitive problems. Research has shown the many benefits of exercise on the body and brain, but even in the absence of intense physical exercise, mental stimulation has benefits and prevents cognitive decline. As long as your cat remains engaged in play, even if they are lazily batting at a toy or lying on their back while grabbing at a feather, it's good for them! I'll discuss more specific techniques for playing with senior cats in chapter 12.

USE IT OR LOSE IT!

A 2017 survey of over 2,600 cat owners found a steady decline in reported playfulness (behaviors such as carrying objects in the mouth, running and jumping, object play, chasing other individuals, chasing lights) as cats aged. Playfulness peaked when cats were one to three years of age, then started gradually decreasing with each passing year, with more notable decreases at ages eight, eleven, and fourteen years.

But less activity doesn't mean *no* activity. Even older cats, on average, showed signs of playfulness, even if only rarely. I personally strongly adhere to the mantra "Use it or lose it." One of my favorite behavior consultations was helping an elderly client learn how to play successfully with her seventeen-year-old cat, Annie, who lived with her in an assisted-living facility. And my own cat Clarabelle played vigorously at age sixteen, until a few weeks before she died of large cell lymphoma.

During the process of domestication, we have selected cats for some juvenile traits, and further enhanced this tendency by controlling our cats' exposure to reproductive hormones via spay and neuter surgery. In our minds, we'd like our cats to stay forever young. Ironically, they age so much faster than we do that we get to experience their life stages from youthful exuberance to distinguished, if not frail. Although we cannot stop our cats from aging and, sadly, also from dying, we can use play throughout their entire lives to help them age well.

CHAPTER 8

Between Friends?

SOCIAL PLAY IN CATS

I think at some point we've all heard cats described as "solitary," "antisocial," or something similar. It wasn't until recently that people started considering that despite the ongoing collective insistence on describing cats in this way, an awful lot of cats do seem to live together. Perhaps we should look into this?

The average household has about 2.2 cats. Of course, many cats live as the only cat in their domicile, but that also means a lot of cats are living in groups (the scientific term for this gathering of cats is a *clowder*). Now, whether all these groups are living in harmony is a topic beyond the scope of this book. Suffice it to say, we now recognize that cats' social lives are much more complex than we have previously given them credit for.

CATS CAN LIVE IN GROUPS

Most of our knowledge about cat relationships is based on studies of colonies of feral cats, which tend to form when resources (space, food, shelter) are abundant enough to support multiple cats. Feral colonies are matriarchal and primarily consist of a few generations of related females who often share caretaking of the young; male cats disperse when they become adults, and non-related males may visit for mating purposes, or they may settle down in one or, if more free-spirited, a few colonies.

There's been much less research on multi-cat households. A 2020 survey-based study found that about 73 percent of owners with multiple cats reported some signs of conflict between them (e.g., staring, chasing, and hissing). That said, owners also reported more friendly signs between cats than hostile ones. Basically when it comes to cats, we can safely say, "Relationship status: It's complicated."

FELINE FRIENDSHIPS

Cats in free-living colonies can avoid one another when desired, but having that freedom and space also allows us to look at the cats who spend more time with one another than would be expected by random chance. The cats who *choose* to hang out together are the cats who scientists refer to as "preferred associates." I like to just call them "friends."

What do feline friends look like? Aside from spending time

near each other, there are some clear signals that cats like each other:

- Grooming each other, in what is called *allogrooming* (*allo* means "same")
- Allorubbing, or rubbing bodies against each other
- Approaching each other with the tail up
- Lying in physical contact, also known as *pillowing*
- Mutual play behavior

If you see these behaviors between cats who live together in your home, there's a good chance that they are friends. Now let's focus on that last one.

KITTENS AND SOCIAL PLAY

Social play appears around two or three weeks of age and dominates the next several weeks of play-related kitten activities, until kittens are around four months of age, when it starts to wind down significantly. I highly recommend making sure that kittens have a similarly aged playmate during their first four months of life. But the ideal scenario is to adopt two (or more!) littermates together, or adopt unrelated kittens of similar ages at around the same time.

A precaution: Due to their highly active and playful nature, I do not generally recommend bringing home a single kitten or adopting a kitten whose only companion will be an older cat. A single kitten will direct their playful and predatory behavior toward anything that moves, including your hands, legs, eyelashes, and hair. Older cats and kittens are at very different stages of life. A kitten will be both frustrated and frustrating. Kittens will seek out any opportunity for energetic encounters, including tackling your ankles or pouncing on an older cat who is trying to sleep. Although some of these matches will work out just fine, I've seen too many cases where a senior cat becomes stressed and even ill, and I've talked to too many cat owners covered with bites and scratches.

In my opinion, it's also not fair to the kitten, who is just being a kitten, to live life without an appropriate playmate. A young kitten is at the most social period of their life and should have the opportunity to express that behavior—with another kitten. Bonus: Adopting littermates is an opportunity to keep a family together! I did this when I adopted three sister kittens in 2020.

Research shows that kittens who are littermates or who are adopted together when young have a better chance of having a good relationship later in life and are more likely to show friendly behaviors with each other, compared to cats who have not lived together for as long. Bonds between cats that form early, form stronger.

WHAT DOES PLAY BETWEEN ADULT CATS LOOK LIKE?

Sadly, we have very little research to fall back on when it comes to play between adult cats. In a study of adult cats in feral colonies by researchers at the University of Georgia College of Veterinary Medicine, social play was often observed, even when the cats were malnourished. A 2005 survey of cat owners revealed that almost 16 percent reported that their cat chased or played with other cats (however, the study did not report how many cats lived alone versus with other cats, so it's hard to know how many cats actually had the opportunity to play with other cats).

There's no one way that cats play, but it can involve several familiar behaviors from kittenhood such as showing the belly, goose-stepping, arching the back, and pouncing. One key feature of play is inhibition—usually keeping those claws retracted, biting gently, making some tentative gestures, and waiting for their play partner to respond in kind.

A bout of play may include stalking, hiding and rushing, chasing, flopping over, wrestling, biting, bunny-kicking, climbing, alternating interactions with the same toy, and general romping. There are often pauses where one or both cats will roll over, maybe have a face-off or extended period of staring (or resting) before the romping resumes. Sometimes the behaviors appear to occur in a consistent, even rigid, pattern; but often the nature of play is, well, playful: unpredictable, experimental, a bit of testing—what happens if I nibble on your tail? How will you respond if I dive-bomb you from the couch? Cats learn from these experiences, which may shape their willingness to engage in play in the future or may help them be a better playmate next time.

ARE THEY PLAYING OR FIGHTING?

Cat owners are often perplexed by interactions between their cats, and none so much as those that fall on the play-fighting spectrum. Remember that many of the social play behaviors we observe in kittens are similar to those we see in antagonistic events between cats later in life. So, the confusion is at times justified. To further confuse humans, cats' relationships are not one-dimensional. Think of times when you fought with a sibling or a friend in your past; even those we love sometimes irritate us and get on our nerves. Should we be surprised that the same is true of our cats' relationships?

But when trying to assess whether your cats are enjoying a wrestle or whether one cat is bothering the crap out of the other, there are a few important considerations:

- **Do the cats otherwise get along?** Do you see them sitting on the same piece of furniture or showing any signs of friendship? Do they nap in each other's presence? Or do they start "wrestling" the second they spy each other?

- **Is the interaction silent or is there screaming?** Hunting and play are generally quiet activities. Fighting involves a lot of screaming and yowling. The occasional hiss or grumble is completely normal. A horrible shrieking sound is a warning sign that things have taken a negative turn.

- **Is the activity mutual?** When cats are truly playing, they often take turns chasing each other or initiating play. I'm more concerned when the behavior is very one-sided, especially if one cat is very young and the other is quite a bit older. It might still be play (at least as far as one cat is concerned), but it might not be mutual. In other cases, one-sided roughhousing can be a sign of outright hostility or an attempt to intimidate the other cat into giving up certain resources.

- **Is the behavior inhibited?** Cat play can get rough—very rough. Put two young male cats together and they will bunny-kick, bite, and chase each other until the cows come home, then take a long nap in each other's arms. But with play, we do tend to see a certain level of "holding back": with bites or scratching, they're not usually giving it their very best, as the purpose is *not* to hurt each other.

- **Are the cats distractible?** If you've ever tried to break up a cat fight, you know that it's next to impossible and something akin to putting your hand in a running blender and holding it there. When cats are playing, usually dropping a magazine on the floor or tossing a few Ping-Pong balls in the opposite direction is enough to get them to stop what they are doing and assess whether something more exciting just happened. One cat may simply walk away from the interaction. Cats who are playing will also typically calm down quickly afterward; cats who are fighting can stay aroused for several hours.

- **Are there injuries?** Cats who are friends should not be giving each other serious injuries. If you are seeing bite wounds, ear notches, or a lot of scabs, things might be a little bit too rough to fairly call it play.

To truly be social play, it takes two to tango. It's important to consider the intentions and feelings of both cats. It is completely within the realm of possibility that to one cat, this play thing is all just good clean fun. Young cats are very playful, often very social, and would love nothing more than to wrestle all day. In my experience, they're often not very respectful of other cats' "personal space bubbles." They are going to bust right through that thing if it suits them.

The other cat may not want to play, or they might not want to play at that moment. They might not enjoy such close contact with other cats. They might have wanted to play a few moments ago, but now they are *done*, or perhaps one cat has crossed a line and now the other cat is downright irritated.

What is play to one cat can be a major stressor to another. Sometimes humans need to intervene by both giving the more playful cat enough interactive play and enrichment to tire them out, *and* giving the more mellow cat some alone time where they can get a break from the rowdy one. Of course, when you have more than two cats in the home, you can do the math and see how relationships and managing them can get complicated quickly. It's important to watch for signs of stress in any of your cats, such as a decreased appetite, hiding, house soiling, overgrooming, and vomiting or diarrhea. If you're seeing these signs and you're confused about whether your cats are getting along, talk to your veterinarian or a behavior professional so that you can know for sure.

HELPING CATS GET ALONG

We can facilitate feline relationships within our homes by remembering the same principle that allows feral cats to congregate in the outdoors: abundance.

We often think of friendship and love as about things we share—the last cookie in the bag that gets broken in half; we make room for our loved ones instead of hogging the couch; we adjust our umbrella when it's raining to cover a friend too—but for cats, sharing isn't necessarily a sign of love. It's not that they are selfish, per se. It's just that everything is so much better for them when they have the choice to share, or more often, to *not* share.

By providing your cats with multiple and separated resources, you help ease the tensions that can arise over competition. Be sure to provide your cats with several food and water dishes, litter boxes, scratching posts, beds, toys, and options for climbing (at least as many of each as you have cats)—and spread them out around your house! This gives cats safe access to the important things they need and gives them options to avoid crossing another cat's path if they aren't comfortable. It means they don't have to share if they don't want to, and they don't have to wait. If you've lived with several roommates and just one bathroom, I think you know what I mean.

PLAYING LIKE CATS AND DOGS

Although we don't have enough research to say a lot about how cats might play differently with other species, cats commonly form relationships with dogs and other pets, not to mention their relationships with the big apes who take care of them, us humans. Let's take a closer look.

About a third of folks with pets are mixing it up by having both cats and dogs. There's no reason the two species can't be the best of friends when introduced and managed properly; in fact, I've known many cats who preferred the company of dogs to that of other cats. There are very few studies that have looked at the cat-dog relationship, and the quality of this relationship appears to be heavily influenced by the cat's comfort with the whole thing. If the cat isn't that into it, harmony may be out of the dog's paws.

Much like you may need to supervise interactions between two cats, you need to do the same when you have a cat and dog under the same roof, especially when they are first getting to know each other. Cats and dogs each have different play styles, modes of communication, and exercise needs. They may or may not be able to figure out how to play together. Your job is to promote friendship and also make sure that each animal is having their play needs met outside their relationship with each other.

I've known a few dogs who have received a hearty smack on the nose from a territorial cat, and sadly, a few cats who have been badly injured or even killed by a very predatory dog. These cases are the exception to the rule, but it's important to recognize what *could* happen and make sure it doesn't.

- At first, supervise, supervise, supervise.
- Always make sure the cat has a dog-free zone to retreat to.
- Give the cat vertical space, where they can observe the dog from up high.
- Give both pets treats, play with a favorite toy, or other good things when they are together.
- Interactions should be controlled at first: use baby gates and have your dog on a lead as they are getting to know each other.
- Give both animals plenty of exercise and play—*separately*! Tired cats and dogs are happy cats and dogs.
- Work with a behavior professional if you're not comfortable with how they are getting along.
- Know the warning signs—a cat who is spending all their time hiding under the bed or hissing and swatting the

dog is not comfortable. A dog who is cowering in fear inside the home is scared of the cat. And a dog who is staring and fixated on a cat may see them as prey.

DOGS AT PLAY

The most classic sign of a dog at play is the play bow: the dog will lean forward with their legs outstretched and elbows touching the ground, with the rear end slightly elevated. Their body language should seem loose, and they may start and stop movements quickly. Dogs may bark excitedly when playing, and their tail is often wagging with enthusiasm. Your cat and dog may spend some time mutually wrestling, nipping, and chasing each other. Much like play between two cats, play between a dog and cat should be mutual, and neither pet should be trying to hide or run away.

SMALL ANIMALS

As much as I would love to have pet rats, guinea pigs, or other small animals, I have accepted that it's not fair to bring an animal into my home that is natural prey to my cats. We cannot train away predatory behaviors in our cats. If you would like to live with a bird, rodent, or similar pet in a house with cats, recognize that it will never be safe to let the two "play" together, because your cat will spend every waking moment plotting to kill them. Keep the "pocket pet" in a room that is off-limits to the cat so that the small animal doesn't have to spend their life being stalked, and make sure they have a both enriching and very secure housing arrangement.

The most notable exception to the warnings against housing cats with animals who may be natural prey is rabbits, who with careful introductions and management can develop strong relationships with cats that even include some playtime. Regardless, it is important to always keep the predator-prey dynamic in mind, supervising closely to make sure that both pets are happy with the arrangement! The House Rabbit Society (rabbit.org) is one of the best resources on taking excellent care of pet rabbits, including how to properly introduce a cat and a rabbit.

BIG ANIMALS: HUMANS

Our relationship with cats is complex; some of us see ourselves as "pet parents," others may see interactions with their cats as more of a friendship, or even a utilitarian, mutually agreeable arrangement ("pest control plus"). The responsibilities of having a cat in your life include providing basic care, such as food, water, a clean litter box, and trips to the veterinarian. For many of us, we hope the interactions also involve lots of affection, such as cuddles and pets. But for too many cat owners, that's where the sense of responsibility starts to wane. It appears that dog owners understand the importance of walking and exercising their dogs (although to be fair, research suggests that about 40 percent of dog owners do not walk their dogs regularly), but there's less consideration for cats' physical needs and how humans should provide for them.

Cats are often described as "low maintenance" for no good reason except circling back to that stereotype of being antisocial. Myths about cats being untrainable, and the fact that cats can use a litter box instead of having to be taken out to eliminate, further contribute to this feline fallacy. I've always felt that you get out of your relationship with your cat what you put into it. You can just give them the basics, but will they thrive? Why aim for such a low bar when we can do so much better? Part of doing better is recognizing that play and exercise are just as important to our cats as food, water, and a clean place to go to the bathroom.

WHAT *DO* CATS DO FOR FUN?

One survey found that when cat owners were asked "What does your cat do for fun?" given the choices: "play with toys," "play with me," "be petted," or "other," 42.2 percent said their cat liked to play with them (e.g., wrestling, laser pointer, Cat Dancer toy), and 83.3 percent said their cat liked to play with toys (e.g., toy mice, balls, bottle caps). A more recent survey out of the Netherlands found that only 40 percent of cat owners said they played with their cats daily, suggesting again that less than half of cat owners are engaging in regular cat play.

Another survey of 283 cat owners attending a veterinary clinic for anything other than a behavior problem found that a larger number (90 percent) of owners claimed to play with their cat at least once a day. A quarter of respondents said they played with their cat for ten minutes at a time, with another third playing for five minutes per session. That said, most cat owners used toys like fuzzy mice, balls, catnip toys, and stuffed animals, and 78 percent of owners said they left the toys out all the time for their cats. Fewer than half of cat owners said they used fishing pole or string toys, suggesting that the owner's definition of playing with their cat may have been a bit loose, and not necessarily the type of play that I'm focusing on throughout this book.

To that end, I really don't believe that 90 percent of cat owners engage in interactive play with their cat every day. This high number contradicts the findings of other studies, and to be included in this survey, cats who were not provided with any toys or activities were excluded, meaning the sample was likely skewed toward cats who received at least *some* play. There may be a social desirability bias going on here as well—where people answer questions in a way that they think will be more favorably viewed (people seem to do this even when surveys are anonymous!)—leading participants to exaggerate the amount of play their cat received.

Although cat owners might be a bit lazy about playing with their cats, they seem to really like playful cats! A 2016 survey of cat adopters found that 68 percent of respondents ranked playfulness as a very important quality in a cat who they were hoping to adopt, far and away rated as more important than characteristics related to looks, being smart or confident, or how well the cat got along with children or other pets. However, we're still left with a large gap in understanding of what exactly it is about playful cats that humans like, since they don't seem to play with them every day.

CAN WE BE A SOCIAL PLAYMATE TO CATS?

Based on how cats behave around us humans, they seem to sense we have the upper hand. They rub on us to solicit attention (and food), much like kittens do to their mothers, female cats do to male cats, and smaller cats do to bigger cats. It's a way to say, "I recognize that you are more powerful than me, and I am letting you know that I trust you." Because we hold this power, we should use it wisely.

We aren't cats, and we should not think we can replace other cats when it comes to play. The kind of play that cats tend to do with one another is more along the "practicing for fighting" spectrum than the refining hunting side of things. To that end, I recommend always using a toy that resembles prey to play with your cat. You are providing your cat with a hunting, not fighting, experience.

Sometimes the lines get blurred. People think it's cute to wiggle their fingers or toes at a kitten, letting the kitten jump on their hands and bite their feet. Other people like to roughhouse or wrestle with their cats. Unfortunately, encouraging this type of play with your cat can set a dangerous precedent—it tells your cat it is okay to interact with humans in a manner that can be harmful. (Literally! Cat bites, especially to the hand, can lead to infections, long-term injuries, and yes, even hospitalization.)

People might think it's cute to play rough with a kitten, but then it's not so cute when the cat is grown up, fourteen pounds, and ambushing them. Other people are okay with it, but their new spouse isn't. Or it's okay until someone has children, and now they worry that the cat will bite their baby. You can imagine who ends up homeless in this type of scenario (hint: it's not the baby). A cat who is surrendered to an animal shelter for biting or scratching has a smaller chance of getting out of that shelter alive than a cat who doesn't respond aggressively to hands approaching.

From the cat's perspective, things aren't necessarily that great either. To be certain, some cats do respond to this type of play as if they enjoy it—but usually because it's what they know, and it's often the only kind of play they are receiving. Other cats can become less certain about humans (remember we are big and powerful and could easily be threatening); should they be afraid of us? Should they be defensive toward us? Should they make the first aggressive move before we do?

Fight and flight have similar functions—they are strategies to protect yourself against threat. We don't want to mix threat and love. We also can't speak cat—we can't tell them in a clear way that they have crossed a line, bitten too hard, or need to back off. And we probably won't realize when they are trying to tell us the same. Most human attempts to correct their cat for biting and scratching, such as yelling, don't make things better. It's much easier to raise your cat to think that toys are for biting and pouncing on, and hands are for gentle interactions and love.

To that end, here are my tips for keeping your relationship with your cat safe and sane:

- Don't mix playing and petting—and don't pet your cat when they are clearly in the mood for play.
- Always use toys when playing with your cat.
- Avoid playing with your cat with your hands or other body parts.
- Avoid chase games (such as peekaboo), which encourage your cat to see you as a threat or prey.
- Don't use those stupid mittens with toys dangling off the fingers.
- If you are playing with small toys with your cat, toss the toys for your cat to chase, rather than holding them with your hand.
- Everyone in the home needs to be on board and treat the cat consistently!

However, if your cat has already picked up some bad habits, I'll be telling you how to work with the play-aggressive cat in chapter 11!

YOUR TURN!

Although I do not support wrestling with your cat, many cats enjoy the physical aspects of that type of play—we just don't want to direct it toward your hands. You can buy a *kick toy* (a larger, tube-shaped soft toy) that your cat can grab and bite while bunny-kicking with their back legs. Or for a nice DIY substitute, just roll up two old tube socks (and why not put a little catnip or silvervine inside if your cat enjoys that) and offer it to your cat. It's a great substitute for cats who may have been encouraged in the past to roughhouse!

Circling back to the original question: Can we be a social playmate to our cats? The answer is yes, but not in the same way as another cat (or even a dog) would be, and certainly not in the way humans play socially with dogs. We should see our role as that of a facilitator of a hunting experience. "I am not a bird, but I am going to move this object for you in such a way that you will be convinced that it is a bird!" We are a puppeteer, a one-person band, a life coach, and a personal trainer all wrapped up in the guise of a human holding a stick with a feather at the end.

SMALLER BIG ANIMALS: CHILDREN

Children are often fascinated by animals—and many of us had our lives changed by experiences we had with pets when we were young. A whole book could be dedicated to helping cats and children coexist happily, but here are the most important elements.

- **Model empathy and good play behavior**: Show your children how to treat cats with respect and how to play with them gently. Help children understand how their cat might be feeling, when their cat might be feeling scared or irritated, and what makes their cat happy. Demonstrate gentle petting, the way your cat enjoys being handled, and avoid rough handling. Your kids will follow your lead.
- **Supervise, supervise, supervise**: Young children should be supervised closely when interacting with cats. When kids are excited, they may grab, shriek, or inadvertently corner a

cat, which can make a cat feel threatened and could result in a bite or scratch. Prevent this situation by keeping a close eye on their interactions until your child is a little older.

- **Set up the home so your cat feels safe:** If a cat is uncertain about an interaction, it's always best for them to have the option to escape. Provide vertical options such as cat trees and shelving, and make sure your cat has safe spaces to get away from activity when they choose. Make sure your kids know not to bother a cat when they are eating, drinking, sleeping, or using the litter box—those are hands-off times!

GROUND RULES FOR SUCCESSFUL CHILD-CAT PLAY

There's no reason that older children can't follow the guidelines in this book and enjoy playtime with cats. In fact, I think some kids and teens enjoy playing with cats more than their parents do. Your children can be a very important tool in your cat-play toolbox!

With younger kids, you will have to provide more hands-on guidance. I recommend starting with softer toys at first, in case your child accidentally bops your cat with a wand toy. It's important to match the toys to your child's motor skills. With younger children, you will likely need to physically guide them by holding the toy and their hands to make sure the play stays gentle. Give your kids praise for positive interactions and point out how your cat is responding to their play, especially when it's going well! Just make sure they can move the toy with enough control not to scare or hurt the cat by accident. Younger children may also be able to gently toss kibble or treats for cats to chase, which for some cats is a great way to make friends.

ARE WE JUST THE WIZARD BEHIND THE CURTAIN?

Cats sometimes seem to notice that people are holding the toy—whether a laser pointer or feather wand. A connection is made from toy to hand. They might see you walk to the toy closet or drawer, anticipating that play will follow. But is this a problem? Not really.

If your cat sees your arm as part of the toy, use a toy with a longer handle, or try wearing a glove while you play to make your hand less obvious to your cat. In most cases, once the toy starts moving in a way that engages your cat's predatory instincts, they won't be so concerned about your hand; they'll be chasing the toy!

IT TAKES TWO

Cats, like us, are individuals. When we play with them, we each bring something to the play space. We can assume that how some cats play will depend on who they are playing with and their relationship with that individual. For example, a cat who was not socialized or is not experienced with children or men (two common scenarios) might need more time to get to know a child or man before being comfortable enough to play with them. A cat might have a long history of playing with a particular person in the home and have learned that play with someone else is much less likely. They might feel less inhibited with some people and get too excited or worked up in play. Other cats will be happy to play with anyone who moves a toy for them. Context matters!

If a cat doesn't play with you—it might be you. It might not be that they aren't a playful cat, but they might not respond to your play style, or maybe the way you play is intimidating or, as much as I hate to say it, even boring. That's okay, I'm here to help. You can keep being you, just a better you who has more kitty-play know-how. But first, let's take a closer look at who your cat is and how knowing that can help you play with them.

CHAPTER 9

Know Your Cat

WHAT KIND OF PLAYER ARE THEY?

To play effectively, you need to know the cat you are playing with. Much like human athletes, cats' sporting abilities can be dependent on life stage, body type, and even personality.

This chapter will help you understand your cat as the precious, unique individual they are before you bust out the cat toys. It will serve as a guide to identifying your cat's personality, which is one piece of the puzzle in understanding what techniques will likely work best and how the different life stages contribute to physical abilities, hunting prowess, stamina, and more. By considering whether a cat is young or old, slender or stocky, timid or bold, and what their hunting style is, we can tailor our approach to play to meet every cat's needs.

CAT'S GOT PERSONALITY

I think we can all agree that our cats have personality; that is, they have consistent traits and tendencies that make them who they are, and that make Fluffy different from Felix.[1] Although there's no "Meowers-Briggs" test, there's been a recent swell of research about feline personality. We can apply this body of knowledge to assess our cats along a spectrum of a few broad categories (such as timid, bold, active, affiliative), which can help us better understand our cats and how they face the world, and how they might face playtime.

WHAT DOES ANIMAL PERSONALITY LOOK LIKE?

As mentioned before, the first key to animal personality is acknowledging that they have individual differences. Personality has been demonstrated in species from honeybees and fish to dogs and cats to primates. These individual differences tend to be consistent across time and different situations. For example, a bee that is more active than other bees today will likely be more active than other bees a month from now (to be fair, honeybees only live one to two months on average) and will be more active than other bees both when in the hive and when foraging.

....

[1] *When referring to animals, some researchers use the term* temperament *instead of* personality*; temperament is sometimes used to refer to innate, inherited tendencies, whereas personality is often described as the consistent ways that individuals feel, think, and behave. I use the terms interchangeably, and there's no clear consensus among scientists that there's a real difference!*

Animal personalities are usually measured using some sort of behavioral test or by observing the animal in different situations, or often by surveying a human who regularly interacts with the animal (such as a cat's owner or a keeper at a zoo). In either case, the measurement is only as good as a measure. If you ask bad questions, have inconsistent methods of observing behavior, or aren't looking for behaviors that are related to the personality trait of interest, then you aren't going to learn much about that individual. Often researchers will compare ratings done by multiple individuals of the same animal to ensure that what one person thinks is bold or aggressive is what someone else agrees is bold or aggressive.

There are some aspects of personality that appear to be common, if not universal, across species (keeping in mind that they may not look exactly the same in different species). In humans, we often describe personality using a model called the Big Five. The Big Five is represented by the acronym OCEAN, which stands for the five personality dimensions: Openness (openness to new ideas and artistic tendencies); Conscientiousness (rule following and orderliness); Extraversion (how outgoing you are); Agreeableness (how easy you are to get along with); and Negative emotions (tendencies to experience anxiety, depression, and the like). It turns out that many of these traits can be found in animals in one form or another. In animals, OCEAN might look like playfulness (Openness); dependability and cleanliness (Conscientiousness); boldness and energy (Extraversion); sociability (Agreeableness); and fear or aggression (Negative emotions).

Although we can imagine different ways that OCEAN might be applied to animal personality, the most common traits that are examined in animals are: boldness, activeness, sociability (or friendliness), and aggressiveness. And truthfully, these four traits are probably more than sufficient to identify many of the differences between our cats. However, researchers have conducted several surveys of owners, searching for evidence that other stable differences exist. Unfortunately, sometimes these studies are fraught with very anthropomorphic concepts, such as describing cats as "proud" or "faithful" or, even worse, using terms such as "mean" or "greedy." Other studies have focused heavily on an outdated term—"dominance"—to try to describe cats. A recent study even suggested that some perfectly natural cat behaviors, such as playing with prey or enjoying perching in high spaces, somehow qualified cats as having "psychopathic" traits. Obviously, using terms like these does more harm than good for cats.

When tempted to describe your cat with terms that are really focused on how humans think and feel and experience the world, take a pause. Do those terms really make sense to apply to cats? How might things be different from a cat's perspective?

WHAT'S WRONG WITH DOMINANCE?

Humans love hierarchies and status; we are naturally competitive. We often describe animals who are assertive as dominant, and I often hear cats called "the alpha"—but what do we mean when we say that? Usually that the cat doesn't follow human "rules" or do what they are told—which is the fault of poor human training technique rather than bad cat behavior!

In the scientific literature, "dominance" refers to the ability to preferentially access resources through force or a natural pecking order that is consistent. But cats' relationships do not follow this structure—first of all, cats do not form "packs" with a strict hierarchy—there are no true "alpha" cats. In homes and in the wild, access to resources primarily depends upon their availability. Lots of mice means plenty to go around and no need to fight. In the home, if you only have one food dish or litter box for several cats to share, it's easy to imagine some tension between the cats. When you have several food dishes or litter boxes spread around the home, conflict between cats goes away. There's no reason to control access to important things when there is plenty to go around!

Furthermore, research in several species has demonstrated that being "dominant" is not a personality trait. It's not about the individual; it's a dynamic between individuals, which again can shift with changes in the environment. Animals that are on the top of the heap in one group of individuals may be on the bottom in another group.

So, I prefer to stick to the most common and clearly

identified traits when thinking about our cats' personalities, and when trying to think about their behavior and motivation; it's always easiest to describe what you see. Instead of "alpha," say "The cat chases other cats when they approach the cat tree"; instead of "dominant," try "The cat sits in the hallway when other cats try to walk by." Taking this approach will help you better understand your cat, perhaps better allowing you to see things from their perspective, rather than assuming that they feel or think about things the same way humans do!

WHERE DOES FELINE PERSONALITY COME FROM?

It's an age-old question: Is personality based on nature or nurture? Well, it turns out you can't blame one or the other for your problems (or positives)—both have an important influence. Research has shown that the same is true for our cats.

A classic study from 1995 found a "friendly father" effect: Even though kittens were raised without their fathers (as is typical for domestic cats), kittens from two specific toms were overall much friendlier than the kittens of the other three fathers in the study. And one male cat in particular was responsible for fathering the least social kittens of the bunch. The effect of genetics is further supported by another study that found that littermates were similar to one another in certain tendencies, such as boldness and how much they liked to rub on humans.

However, early handling was also very important, with kittens who received more handling before eight weeks of age showing more boldness with humans when they were four months old. As I'll address later in the book (chapter 11), when discussing why some cats are fearful, kittens have a short but very influential socialization window during which handling and positive exposure to people, places, and things can help them be more easygoing, confident adult cats.

Finally, the environment a cat currently lives in also contributes to how they behave and how they express their personality. For example, a shy cat is unlikely to thrive in a very loud, active household with lots of people coming and going. They are going to become even more fearful and withdrawn. In a quiet, calm household, they may still be shy compared to other cats, but their owners might forget they have a shy cat until the doorbell rings or houseguests arrive! An active cat may thrive in a home where they get lots of playtime and exercise but become an absolute terror in a home that is understimulating to them.

Although personality is a relatively stable thing, it is not an absolute. The seeds are sown (literally) early in life through the genetic hand a cat is dealt. Their early-life socialization molds that clay into one shape or another. But the clay that is personality may also shift over time, as your cat has a variety of experiences, learning opportunities, and hopefully an environment that allows them to show their best side.

GETTING TO KNOW YOUR CAT

Think about the major facets of animal personality, and think about where your cat might fall on the spectrum related to each of those traits:

SUPER TIMID AND SHY ———— OUTGOING AND CONFIDENT

LAZY AND MELLOW ———— HYPER AND ACTIVE

INDEPENDENT ———— VERY SOCIAL AND INTERACTIVE

GENTLE ———— WILD CLIFF DIVER

PERSONALITY TYPES

MELLOW

BONKERS

HOW DOES PERSONALITY IMPACT PLAY?

Now take a moment to think about who your cat is and how their personality traits might impact your playtime together.

We've got two goals here:

- To help you understand who your cat is just a little bit more
- To set up your play sessions for success

For example, a shy cat might be more easily startled by large toys or might need more safe spots to hunt from. A less active cat might have a slower warm-up time, and a super-active cat might need longer or more frequent play sessions. If you've got a cat who tends to get a little aggressive in play, you might need to use wand toys with extra-long handles (we'll talk about how to work with those kitties in chapter 11).

PHYSICAL TRAITS

All cats have a blueprint that allows them to hunt successfully. In that sense, all cats are hunters, and therefore all cats can be players. That said, there are some cats who can turn things up to eleven, so to speak. It might be their energy level that allows them to go, go, go, or it might be their athleticism that allows them to do amazing leaps and backflips. They might have long legs that allow them to propel to unknown heights when chasing a feather.

When you're setting up your play area, or planning your play session, or waving that wand around, keep in mind your cat's body type, physical strengths, and limitations (e.g., can they make the jump from the couch to the cat tree? Can they reach where you are waving the toy?). This can help you determine when you can increase the challenge or when you need to make life a little easier for your feline friend.

You'll also want to consider if your cat has any medical conditions (e.g., hip or leg injuries) that might call for a modification of play techniques (such as keeping the play less vertical and more on the ground, while making sure there's carpet or rugs for stability). If your cat has a health problem, such as heart disease, be sure to ask your vet about how much activity is okay for your cat. For example, when my cat Clarabelle was diagnosed with congestive heart failure, her cardiologist said that low-key play was acceptable, as long as we monitored her breathing and let her dictate the pace (she was ten, so the pace was pretty mellow). If your cat is overweight or older, you'll want to start slow. I'll cover how to play with special-needs cats, including cats with disabilities, in chapter 12.

FELINE BODY TYPES

Cats actually come in six body types:

Substantial: These cats are just large! Think the Maine Coon breed.

COBBY TYPE
E.G., MANX

FOREIGN TYPE
E.G., SIAMESE

Cobby: These cats are short and wide due to their build, not due to being overweight. They are compact cats, often with short, strong legs and a round head, as seen in Burmese cats.

Semi-cobby: Not as cobby as cobby cats, but still a little cobby (example: British shorthairs).

Medium: Your average domestic cat; they do not have extreme facial features, and honestly, they look "just right."

Foreign: Long legs and tail with a muscular body. They often have a long face and large ears, such as in the Siamese breed.

Semi-foreign: Similar to the foreign body type but less extreme. They are graceful with long legs and a bit more svelte than medium-built cats, such as in Abyssinian cats.

THE EFFECT OF BREED: GETTING FANCY

Although people are quite charmed by the unique appearances of many "fancy" cat breeds, purebreds make up a relatively small proportion of domestic cats. Cat breeds are lineages of cats who have been selected for specific traits. Although purebred cats are generally bred for looks and not hunting skills or athleticism, they can vary in their enthusiasm and natural proclivities toward exercise and play. Humans have only been creating purebred cats for a relatively short period of time, and biologists estimate that more than 96 percent of cats are "free breeding," meaning that they choose their own sex partners. Purebred cats are still the same species as the "mutts" that most of us have sitting on our couches (domestic cats), and all cats have a shared ancestry.

There's a saying that "breeds have more variability within than between." Let me translate that for you: We recognize that each individual is unique—there are cats who are more outgoing, cats who are vocal, cats who are shy, cats who are rambunctious. The same is true within cat breeds. Even among Siamese, some will be more vocal than others, some will be more outgoing than others, some will be cuddlier. We also don't expect every Siamese cat to be the same! When we compare breeds of cats to one another, what we see is that it's hard to find large and consistent differences in personality between breeds because during the process of breeding cats, there has been more focus on selection for appearance than personality.

That said, there have been a few studies suggesting that breed may have an influence on playfulness and activity—at least in some breeds. For example, in a 2017 survey of personality traits in domestic cats, cat owners reported Bengal, Burmese, and Siamese cats to be the most active, with Persians and Ragdolls ranked as less active. It's possible that by choosing a specific breed of cat, you can find a cat that matches your own personal play interests. (Love to play? Maybe a Bengal is for you. Lazy? A Ragdoll might be more your speed.) But even if you get the raggiest rag doll of a Ragdoll, you're still going to need to play! And remember that due to that variability within a breed, some of those Ragdolls are still going to be very rambunctious.

Our understanding of feline genetics has advanced in recent years, so perhaps the cats of the future will be bred just as much for personality as looks, meaning that we could ramp up or tone down their hunting instincts and playfulness. But for now, we must recognize that genetics are just one small piece of what makes a cat who they are: how they are raised and how healthy an environment they live in are likely *just* as important.

DOES THE SEX OF YOUR CAT MATTER?

Are male and female cats different in their play styles? It's an interesting question, and it's not unreasonable to think there might be differences based on hormonal influences. But much like the effects of breed are minimal, a kitten's behavior is much more likely to be influenced by genetics, early-life socialization, and the environment than by their sex. By spaying and neutering, we may also be reducing the hormonal effects, especially in adult cats. That said, a few researchers have tried to tease apart how male and female kittens might differ when it comes to play.

One study found no sex differences in play behavior, but female kittens were overall more active than male kittens. However, multiple studies have found that at least in kittens, males make more frequent contact with toys between six and twelve weeks of age compared to female kittens. Females in a litter with some male kittens also made more toy contacts compared to kittens in all-female litters, and the more males there were in the litter, the larger the difference.

A similar tendency was found in older kittens (twelve to sixteen weeks of age) in measures of social play: male kittens from all-male litters were more likely to engage in social play than all-female litters, and when there were males and females in the litter, female kittens played more when there were more male siblings around. This suggests that exposure to male hormones (androgens) in the womb may have an influence on both social and object play in female kittens. However, male kittens do not seem similarly affected—more female kittens in the litter do not change male play behavior.

When it comes to adult cats, we have much less data. Studies of hunting behavior report no sex differences in frequency of bringing home prey. We might even expect that female cats, when weaning kittens, might have to be a bit more on the ball when it comes to hunting when they've got mouths to feed and kittens to train to hunt (male cats do not participate in the care of their young)—but at least in well-fed house cats who aren't raising young, we do not see such differences between the sexes. Finally, a 2017 survey out of Australia also found no sex differences in reported playfulness (i.e., cats identified as energetic, playful, quick, mischievous, and curious) in adult cats. You can rest assured, regardless of whether you have male or female cats, there's no reason to expect that this will have a huge impact on your playtime—you'll want to assess your cat's play interests individually and adjust accordingly.

ALL WALKS OF LIFE

Cats who are indoors only tend to be more active during the day than at night, and their activity patterns are heavily influenced by their humans. In other words, if you're around and awake and interacting with your cat, they're more likely to be active. But when you're at work all day, your cat is likely doing a lot of snoozing. Cats with freer outdoor access, especially overnight, are more likely to adjust their schedules to a lifestyle more akin to that of a vampire.

Now, you might be thinking that, deprived of the freedom and the "all-you-can-kill" buffet lifestyle of cats who go outdoors, your indoor cat might be less inclined to play. Do they even still have hunting instincts if they don't get to practice?

Not so fast! A 2021 study tested the responses of both indoor-only and indoor-outdoor cats to different types of toys (a wand toy and balls), as well as bird, mouse, and rustling sounds. They found that indoor-only cats were quicker to interact with the ball toys and responded more quickly to the prey-like sounds. Is it because cats who go outdoors cannot be fooled by our fake prey offerings? Or are indoor-only cats so deprived that they'll play with anything that moves or squeaks? Only more research will tell, but I'll address playing with cats who have outdoor access in chapter 12.

CHAPTER 10

Four Important Lessons about Cat Play from Science

Let's take time to review some of the key research and scientific thinking that can further enhance our play technique. Thankfully, scientists have explored a few important play questions, such as: What do kittens like to play with? Do cats get bored of the same toy day after day? Does it matter if the cat toys look like prey? And does my cat prefer me or the toy?

LESSON 1: WHAT TOYS DO KITTENS LIKE TO PLAY WITH?

A 1977 study called "Exploration and Play with Objects in Domestic Kittens" tested the behaviors of twelve kittens (six male, six female) who were living in a laboratory with their mother. The purpose of the study was to assess kittens' responses to familiar and unfamiliar items, investigating exploratory behavior (looking at, licking, sniffing, and/or touching without manipulation), play (tossing, grasping, and mouthing of the object, often preceded by pouncing or leaping on the object), and exploratory play (exploration of an object always preceded by running or leaping over, around, or into an object).

Twice a day for five days, kittens were brought individually into a large room, where they were presented with one of ten different items they had never been exposed to before: a wine cork, a small cardboard cylinder (the scientific term for a toilet paper roll), a paper bag, a shoebox, a shoe, a suspended piece of string, a rubber mouse, a ball of yarn, a Ping-Pong ball, and, finally, a dead mouse. Each item was presented once, and each kitten had their own set of items so they wouldn't smell like any of their littermates.

After each testing session, the item was placed in the kittens' regular housing (with Mom), aside from the dead mice, which were replaced with "fresh" dead mice every day. On the next five days of testing, the kittens were retested with identical items, in the same procedure and sequence as the first five days. The thinking was that the first time each kitten encountered each object, it was novel and exciting. The second time around, it was familiar since the kitten had been living with it for five days. Would this influence kitten behavior? And which of the objects presented would they prefer?

The kittens spent more time exploring and less time playing

the first time they encountered an object. The second time, the opposite was true—they spent less time exploring and more time playing. Kittens also spent more time "hanging out" in the second session compared to the first, and they were less exploratory and active overall.

Perhaps not surprisingly, the kittens' favorite items to explore and make non-play-related contact with (i.e., things to sleep in or on) were shoes, boxes, and paper bags (demonstrating that even in the 1970s, cats were cats). Although the purpose of the study was not to see how much time kittens spent sitting on things, this is what seemed to increase with familiarity. Kitten play was most prevalent with the smaller objects, such as the cylinder, the Ping-Pong ball, the string, the cork, and . . . the dead mouse. The overall lesson of this story was that the kittens preferred smaller items that were easy to manipulate, and they seemed more willing to play with items that were slightly familiar.

TAKEAWAYS:

- Kittens prefer small objects that are easy to move around.
- Kittens need a little time to get to know their toys.
- True to form, kittens love boxes and bags.

YOUR TURN!

It's not just kittens who love boxes and bags—adult cats do too. Try placing a box or paper bag (with any handles cut off for safety) on the floor at the start of your next play session. Most cats can't resist investigating! Allow them to use the box or bag to hide in while they stalk a toy, or move the toy in, under, or around the box or bag to entice your cat. When the play session is over, they may even take a nap in it!

LESSON 2: DO CATS GET BORED OF THEIR TOYS?

In light of the previous study, which suggested that kittens might like *familiar* toys, we should talk about whether the same is true of adult cats. People often mention that their cat seems bored of their toys, and this is probably because your cat *is* bored of their toys, especially if you use the same toy over and over or even for too long during a single play session.

Habituation is the decreased response to a stimulus when we are exposed to it repeatedly or for a prolonged period. For example, when we first got a fountain for our cats, the splashing noise of the moving water was very prominent and distracting to me; now I barely notice the sounds. We don't give a second thought to the sensations of familiar clothing on our skin, because we have habituated to it (but if you put on clothing made of a new, rough material, you would certainly notice at first—and then once again habituate).

Habituation occurs across the animal kingdom and even happens when animals are doing things that should be fun. Activities that are enriching can be rewarding on their own—behaviors like hunting and exploration are necessary for survival, and animals will do these things even without an immediate reward (for example, the several hunting attempts that are unsuccessful do not stop cats from trying again!).

But if you are primarily motivated to explore something in your environment (say a new toy) because it is new, as the novelty wears off, so will your interest. Without an additional benefit (such as a food reward) to reinforce the investigation of the new object, soon it will be left neglected. This is one reason that people observe their own cats losing interest in toys—because they are no longer novel.

But for every habituation, there can be dishabituation—the "rebound" of interest or response to a stimulus at some point after you habituated. This usually occurs after something in the environment has changed, or when the stimulus to which you habituated disappears for a while. When it comes back, you may have a similar or even bigger response than you did the first time. For example, maybe your partner is baking chocolate chip cookies. You walk into the kitchen and are greeted with the amazing scent combination of vanilla and chocolate. As you get yourself a glass of water and chat with your partner, you stop smelling the cookies. You have habituated. You leave the room and go back to your office and do some work. Twenty minutes later, when you return to the kitchen, the smell of chocolate chip cookies is even more powerful and enticing than it was before (and luckily, the cookies are ready to come out of the oven).

To further understand this phenomenon, Dr. John Bradshaw and Dr. Sarah Hall used a very clever study to determine whether cats habituate to (get bored of) toys. The first two experiments were done with group-housed adult cats who had never hunted anything bigger than a spider. Cats were chosen for playfulness and confidence, as they were tested individually in a new room.

In experiment one, eight cats were tested by an experimenter who swung a string with a small furry object attached to the end. Each cat was tested with the same toy three times in a row, and on the final trial, they were presented with either a new toy or with the same toy from trials one to three. Each cat was given the opportunity to interact with the toy for two minutes with a twenty-five-minute break between trials. The cats were given four test sessions total, with at least a day's rest between sessions.

The researchers measured how often the cats clutched or bit the toy on each trial. By the time the cats got to the third trial with the same toy, they were already biting and clutching it much less (*bor-ing!*). When the toy was not changed on the last trial, most of the cats were no longer interested in playing. When the toy *was* changed on the last trial, the cats played with the toy more than they did on the third trial, although not quite as much as they had on the first trial. Dishabituation in action!

Now that it was demonstrated that the cats were more interested

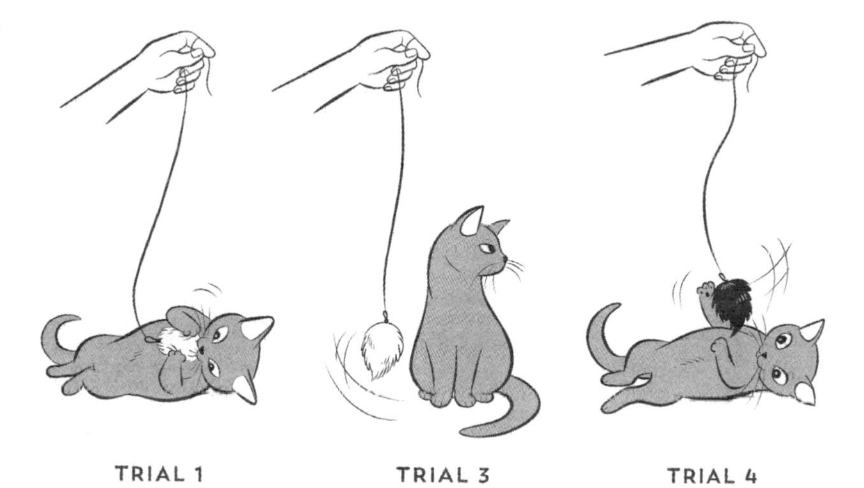

TRIAL 1 TRIAL 3 TRIAL 4

in a new toy by the fourth trial, the researchers next tested whether it mattered how long cats had to wait between trials. Experiment two was very similar to experiment one, except in this case, the researchers varied the time between trials to either five minutes, fifteen minutes, or forty-five minutes. The results suggested that a long break between play bouts did nobody any favors. When the cats had to wait forty-five minutes between play trials, the cats played less in the fourth trial than in the first (no dishabituation). When the cats had to wait only five minutes between play trials, they played *more* in the fourth trial than they had in the first trial, suggesting a much stronger dishabituation effect (the results were middling when cats had a fifteen-minute break).

To confirm what we'll call the five-minute rebound effect, the researchers repeated the study in eight pet cats with a known history of hunting. Much like the cats in the second experiment, the cats who got a new toy in the fourth trial increased their play to a higher level than in the first trial. The cats who got the same toy on all four trials were barely playing by the end—in fact, only one cat played in the fourth trial when the toy wasn't changed.

This study demonstrates how important it is to switch out toys when we are playing—and the same is probably true of most enrichment we provide for our cats, whether it's props for play, interactive toys, solo toys, or activities like cat grass, catnip, food puzzles, or even a cardboard box to explore. When in doubt: rotate, rotate, rotate! A survey-based study of cat owners revealed that most (78 percent) did not rotate any cat toys, instead letting them languish and collect dust under the sofa. Do your cat a favor and pick up all those toys at least once a week (it makes it easier to sweep or vacuum too!) and swap them out for different toys (as this study demonstrated, even just a differently colored version of the same toy may suffice).

TAKEAWAYS:

- Rotate toys frequently, both between and within play sessions.
- You can "refresh" a toy by putting it away for a few weeks, then bringing it back into the rotation.

- Just because your cat is bored of a toy does not mean they are bored of playing.

YOUR TURN!

Next time you play with your cat, try switching to a new toy when your cat's interest starts to wane. Much like in Dr. Bradshaw and Dr. Hall's study, you can also try a five-minute pause before making the switch. Did this renew their interest in your play session?

LESSON 3: IF IT LOOKS LIKE A MOUSE, MOVES LIKE A MOUSE, AND FALLS APART LIKE A MOUSE . . . IS IT A MOUSE?

Is there anything we can do, besides frequently changing out toys, to stop our cats from getting bored of play?

Dr. Sarah Hall proposed two toy features that can prevent or at least forestall boredom. First, cats are sensitive to a prey item's "changeability"—as the prey changes, the cat continues to be excited and will continue to hunt. A prey animal will struggle but weaken, may crunch in the teeth, start bleeding, lose feathers, change physical form, get colder . . . and this is starting to sound like a scene out of *Dexter*. All these changes signal to a cat that they should continue biting, grabbing, and kicking, as they are succeeding in their goal—killing.

Conversely, there may be circumstances when cats *should* habituate to prey. A good reason might be if they are hunting someone who is too strong to succumb to their predatory attack. No bleeding, no lost feathers, no weakening . . . might be signals that it's time to give up. Sorry, kitty, that bird is just too big for you.

It follows that toys that do not break down easily might lead to faster habituation. The message the cat receives is a static signal—prey that cannot be conquered. One way we can keep our cats engaged in play is to offer some of that "changeability" with toys that can safely be shredded, torn apart, or that can stand to lose a feather or two.

Second, Dr. Hall also emphasized the importance of the characteristics of the toy. A toy that is more similar to prey is more likely to signal to a cat: *Hunt me!* Toys that are small, complex in shape or textures, with fur or feathers, and that can be moved quickly (such as by being on the end of a wand) are more likely to attract and maintain a cat's interest.

WHY DOES MY CAT LIKE THIS TOY THAT LOOKS NOTHING LIKE A MOUSE?

Now, it is true that your cat might be wild about that toy that looks like an avocado or an ice cream sundae. Things your cat wouldn't be hunting. So why would a cat play with these toys if it's important for toys to be prey-like?

The best explanation is that there are other factors involved that might trigger your cat's play instincts. For example, if that avocado is stuffed with catnip, that is likely to excite some kitties and get them playing even if the toy isn't very lifelike. Some cats may be more "creative" in the sense that they might be willing to suspend any disbelief and just err on the side of "well, this avocado is fuzzy, so maybe it's actually a mouse." Your cat's threshold for how prey-like a toy has to be might be very low, whereas other cats might actually be quite picky and only play with toys that are very similar to prey. From a survival perspective, there might be times when it's better to assume that if it might be something to kill and eat, it's better to try to kill and eat it than let it get away. At least until proven wrong!

For our pet cats, luckily there's no harm in them preferring that taco toy over the fuzzy mouse. It's another reason to go into play with the mindset of an experimenter and be willing to try a variety of toys with your cat. They just might surprise you with what they love to play with most of all.

TAKEAWAYS:

- Use small toys that are easily moved and manipulated by your cat. The more prey-like, the better!
- Think fur, scales, and feathers—and look for toys with interesting, complex textures, which will likely elicit more predatory behavior than simpler versions.
- Give your cat the opportunity to destroy some of their toys—paper or cardboard, or chewable/edible toys.

- Experiment with different types of toys to see what your cat likes—your cat might surprise you by loving some toys that aren't anything like prey too!

LESSON 4: OKAY, KITTY, IT'S ME OR THE TOY. OR BOTH?

People seem to want a lot of reassurance that their cat loves them. If you search for "Does my cat love me?" on a popular search engine you will get approximately 2.6 billion results back. All cats are unique in how they express affection, and I refer you back to those signs of good welfare on page 96. As much as I would like my cats to love me, it's just as (if not more) important that they are happy. A happy cat is more likely to show their love!

TIPS FOR MAKING YOUR CAT LOVE YOU

1. **Let your cat come to you, and let them walk away when they are done:** Giving your cat a sense of control over their physical interactions with you helps them trust you. Many cats prefer when we play a little hard to get.

2. **Respect cats' preferences when it comes to petting:** Research has shown that *most* cats prefer petting on the cheeks and forehead, and that the base of the tail

is cats' least favorite place for pets. Of course, each cat is an individual, but it's important to keep these preferences in mind as you get to know a cat.

3. Provide your cat with a safe environment they can thrive in: Your cat needs to feel comfortable in their environment to do all the healthy cat things. This means having appropriate structures to climb and scratch, a clean place to go to the bathroom, and safe places to hide and eat.

4. Socialize and train your cat: Cats benefit from gentle handling and positive associations with new experiences and people when they are young. But learning is lifelong, and we can use rewards-based training methods to help our cats have fun, accept routine handling (like nail trims), or be well behaved.

5. Play with your cat!

♥ ♥ ♥

Maybe you're the jealous type and it's not enough to know that your cat loves you. No, you also want to know if your cat loves you *more* than they love other things. Things like tuna and catnip. Well, once again, *science to the rescue*. Dr. Kristyn Vitale dedicated her PhD research to better understanding the human-cat bond, including whether cats prefer human interaction over other activities.

To assess cats' preferences between four types of enrichment (toys, treats, scent, and human interaction), nineteen shelter cats and nineteen house cats were put through a battery of tests, pitting one form of feline entertainment against others. For the two and a half hours before cats were tested, all food and social attention were withheld so they would be hungry for food and love.

In four separate sessions researchers offered cats three types of each enrichment category:

- Toy session: moving toy with feather, a mouse toy, a feather
- Food session: chicken, tuna, chicken-flavored soft treat
- Scent session: gerbil smell, catnip, unfamiliar cat scent
- Human interaction session: human talking to the cat, petting the cat, or interacting with the cat with a toy

Cats had three minutes to investigate the toys, then the toys were taken away, then they had three minutes to investigate the snack choices, and so on. Preferences were measured by recording how long cats spent contacting the enrichment choices, such as touching, eating, sniffing, or playing with them. For the human interaction condition, the same person spent three minutes with the cat, spending one minute engaging in each of the three activities.

Perhaps unsurprisingly, the cats loved catnip and tuna. But in a win for play, two thirds of the cats preferred the human playing over talking or petting, and about the same number of cats preferred the moving toy over the mouse or unanimated feather.

After each individual cat's top choice for each category was determined, those four enrichment types (tailored to each cat's preferences) were offered simultaneously in a head-to-head competition. For three minutes, cats were presented with a four-square, with a toy in one quadrant, a scented item in another, a snack in quadrant three, and a human in quadrant four (ready to talk, pet, or play). Again, researchers measured the time cats spent interacting with each type of enrichment. Would cats prefer catnip, tuna, playing with a human, or a moving toy?

YOUR TURN!

Can you reproduce Dr. Vitale's experiment? Try offering your cat a simultaneous choice of a toy, a catnip-scented item, a treat, and you! Which does your cat investigate first? Are the results the same if you try again on a different day?

When these were placed head-to-head against one another, cats showed a strong preference for social interaction. Nineteen cats chose the social condition, whereas just four chose the moving toy without a human attached. But since most cats preferred the human + toy social condition, we can infer that the combination is irresistible to many cats—science says so!

TAKEAWAYS:

- Each cat is individual in their preferences, but preferences may be temporary.
- What your cat wants right now isn't necessarily what they will want in five minutes.
- Cats like humans, and they like moving toys, but they may like the combination of the two best of all.
- Play is just one way to make your cat love you more.

CHAPTER 11

Play As Therapy

HELPING COMPLICATED CATS

Working in an animal shelter made me see the light—that playing with cats could *help* them. I watched incredibly shy cats come out of hiding to bat at a Cat Dancer toy and eventually become more confident so they could meet potential adopters. I worked with cats who were frustrated at being cage-bound and observed how they were able to breathe a sigh of relaxation after an intense play session. Cats who bit or scratched humans could have that energy diverted to a feather teaser instead of a hand. All these cats became more adoptable through play.

But play isn't just for stressed-out cats in animal shelters; it can do wonders in the home too. In the past twenty years of my work with cats and their humans, play has become one of the *most* important tools I use to help my clients prevent and solve behavior problems and to improve overall cat welfare. It's hard for me to think of very many cases in which I didn't recommend that people play with their cats more (often because they were rarely playing with them in the first place).

"FIXING" BEHAVIOR

Perhaps the most important thing I've learned is that changing a cat's behavior is much easier than changing that of their owner. There are various reasons that folks struggle with understanding their cat, but I think one of the biggest issues is that it is difficult to step outside of our own head, our own experience, to try to see things from our cat's perspective. Oh, and the language barrier doesn't help either.

Although many of us are equally terrible at communicating with our human loved ones or coworkers, at least we have a common language to do so. But your cat can't read your mind. They don't understand why you are frustrated, disappointed, or angry. Your body language and tone of voice tell them that something is wrong, but they don't know what is wrong or how to change. A cat's response to this kind of behavior (angry sounds, tension) is often fearful or defensive, usually nothing desirable to humans. We often make things worse by misinterpreting our cat's behavior with an anthropomorphic (human-centered) lens: assuming our cat is spiteful or is doing things to get revenge or to deliberately agitate us, rather than recognizing that our cat responds to their environment, and that stress can lead to feline behaviors that we don't like. Those "bad behaviors" are often a cry for help.

I saw this feline-human relationship meltdown hundreds of times when I worked with homeless cats in the animal shelter. I saw how behavior could fracture a bond and cost a cat their home; behavior could hinder a shy cat from getting noticed by potential adopters or lead a coworker to label a cat as "angry" or "not nice." In many cases, things that we consider behavior problems in our cats are just *normal* responses to an abnormal or otherwise inadequate environment. We expect cats to use a disgusting litter box, we expect them to be content with a boring life, we expect them to live peacefully with other animals they don't feel safe around, we expect them to cuddle when *we* want to, and we expect them to behave as if they are not hunters.

To modify our cats' behavior, we must modify ours first.

With the thousands of clients I have helped over the past twenty years, there was one universal thing I had observed. In almost every single case, the humans were not playing with their cats enough, if at all: "He gets bored of toys," "She's not that interested in playing," "When we have time to play, she's not in the mood," "He gets too excited from play, so we have to stop." For some people, it honestly didn't seem to occur to them that their cat needed daily exercise, or that they should facilitate that exercise. Many people attest to dusty piles of fuzzy mice or Ping-Pong balls sitting in a basket in the corner, not understanding why those toys no longer elicit excitement from their cats.

Although play is just one part of a behavior modification plan, it can be one of the most important ones. While I can't provide a complete guide to fixing all cat behavior problems within these pages (there are several excellent books already out there on solving feline behavior problems, or an experienced and qualified behavior professional can help you!), I wanted to highlight the importance of play and some specific techniques I recommend in some cases.

(Note: For any behavior problem, and especially when there is a sudden change in your cat's behavior, always start with a veterinary visit to rule out any medical conditions.)

TAMING THE WILD BEAST: PLAY AGGRESSION

Nothing leads to more emails of "HELP! My cat is attacking me out of nowhere!" than play aggression, which sounds harmless or even fun. "Play" or "predatory" aggression is when a cat focuses their hunting behaviors toward human body parts, often hands and feet, typically resulting in stalking and ambushing their owner. I encourage people to avoid words like *attack* when describing their cat's behavior, because it's not a description of behavior at all. To some people a cat merely hissing (a form of communication) is scary; other people find it amusing that they are dripping blood after being bitten by their cat. Instead, let's think about what your cat is doing. Biting, scratching, and bunny-kicking are often part of the behavioral presentation. It's no joke when a cat's killing bite is directed toward your ankles, and this behavior can be terrifying or even dangerous (especially if directed at someone's face, or at someone on blood-thinning medications).

Play aggression can be easy to assess because there are usually a few classic signs:

- The cat is young, often living without a similarly aged playmate.
- "Attacks" are almost always silent.
- The cat shows classic "hunting" body language (dilated pupils, alert whiskers, stalking, butt wiggle).
- The cat may hide before pouncing (e.g., under beds or furniture, around corners).
- The cat pounces on or bats at moving body parts—dangling feet or hands, chasing the owner as they move through the home, batting at hair or fluttering eyelashes.
- The behavior may be worse when the cat is hungry, during playtime, or at dawn/dusk.
- The cat may have been encouraged to see hands or feet as toys (e.g., wrestling or wiggling fingers or moving toes under the covers to get them to play as a kitten).
- The owner often reports the cat is easily bored or otherwise gets into trouble; a classic sign that the cat is not getting enough play or mental stimulation.

It's hard to convince a cat owner that, hey, your cat is just playing, but that is what is going on here. It's really nothing personal except that your cat has just one thing on their mind—killing—and right now, you look like a giant, tasty bird or mouse to them. We often think of hunting as reactive in cats, stimulated by a trigger—a sound, detection of motion—that flips a switch. In the case of play aggression, it *feels* a lot more like the switch is flipped first, the cat goes into hunting mode without a clear target. When they are in search of prey, almost anything that moves (your hands, your feet) will do.

When your cat has a serious case of play aggression, your first response should be collecting a little data. What time of day do the pounces occur? What does your cat do? Biting? Scratching?

Pouncing and grabbing? What body parts does your cat target? This information will help you understand when you need to focus on preventive measures, and what kind of prey/play movements they might be most interested in.

WORKING WITH PLAYFUL/PREDATORY AGGRESSION

This is a case where more play is necessary. Sometimes people think the play gets the cat *more* worked up (and to be fair, it can seem that way because the hunting switch may get flipped when the cat gets excited), and so owners tend to stop playing with the cat much, if at all. Rather than having the desired effect of calming the cat down, this creates a kitty pressure cooker where the predatory energy builds up, leading to frustration or even escalation of the problem.

Play-aggressive cats need several play sessions a day, with an appropriate cooldown, where the play is gradually slowed down at the end. Offer play throughout the day so your cat is not snoozing for too long at a stretch. Play sessions can vary in length, but plan on at least ten to fifteen minutes for each, with an appropriate cooldown. Longer sessions may be needed!

YOUR TURN!

The perfect way to end a play session is just like your cat would end a hunt: with a snack. This is also a great way to direct your cat away from your body while you put the toys away. At the end of your next play session, try skittering a few kibbles or dry treats across the floor (far, far away from your body) for your cat to chase and eat. Happy cat, safe human.

Be prepared: Carry small toys (e.g., fuzzy mice, crinkle balls) or interactive wands around the house with you. The goal is to redirect your cat's excessive energy toward something appropriate. You are essentially training your cat that when you walk through the house, toys go flying *away* from your body. Even if they aren't always actively stalking you, it will divert your cat when they are, and this exercise gives them plenty of opportunity to practice pouncing on other moving objects besides your body.

For bunny-kicking, have large, kicker-style toys handy: these are toys that are typically a little larger and in a tube shape that allows your cat to grab with their front legs while kicking with their back legs (you can even make a kick toy by rolling a few old tube socks together). Toss the toy *gently* toward your cat's stomach, allowing them to grab it; avoid holding the toy with your hands while your cat is grabbing and kicking it, lest they see you as part of the prey or accidentally swipe you.

For cats who like to lie down and grab your leg while kicking your foot, you can also give them an opportunity to be prone and kick by taking a cat-scratching post and tipping it on its side. Some cats like to treat the post almost like monkey bars, using all four paws to grab on and move up and down along the post.

Remember: we aren't trying to squash this aggressive behavior—we are giving it the outlet it needs. With time, aging, lots of repetition of diversion, and so much exercise, this behavior does get better. Understanding that your cat isn't mad at you or trying to drive you out of their territory can help; when people feel like their cat is out to get them it can lead them to feel scared or have a large reaction to the behavior.

Most importantly, you must stay calm when bitten or pounced on by a play-aggressive cat. I know this might sound a bit flippant—oh, just stay still while your cat is digging into your arm—but if you scream, yank your arm or foot away, or otherwise have a large physical reaction, you are behaving like prey; such a reaction almost always increases a cat's excitement and can lead to further injury (e.g., biting or kicking harder) or, worse yet, the cat becoming fearful and defensive.

In general, a play-aggressive cat can benefit from other behavior modifications, including clicker training, lots of environmental enrichment, or a similarly energized feline playmate (let your cat bite and kick them instead!). Some cats may need behavior medications, so again, don't hesitate to talk to your veterinarian or a veterinary behaviorist to get help if you've got a panther who is just a little too wild.

FEARFUL CATS

Fearfulness can manifest in behavioral responses to a specific situation (such as hiding when the vacuum turns on or being scared during a trip to the veterinary clinic), or it can be represented by a more diffuse, nervous personality. These are the cats who may run under the bed when the doorbell rings, who your friends never see ("No, really, I swear I have a cat!").

I have the softest spot in my heart for fearful cats; these cats are often mislabeled as or assumed to be abused, because they may shy from hands approaching them or just run away when approached at all—unless they know you very well. Most of them are just what we call undersocialized, meaning they lacked the positive exposure to humans when young that would have made them more comfortable around us.

Cats have a "sensitive period," which is their time as kittens when they are most open to learning new things and can quickly make associations between positive experiences and individuals, be they human, cat, dog, or other. Kittens who are exposed more frequently and to several different people are likely to generalize that positive experience they had with a few people to pretty much all humans. These are the cats who greet you at the door when you visit friends, or run to meet you on the sidewalk when you're strolling in the neighborhood.

Unfortunately, this window is short and early: it happens when kittens are approximately two to seven weeks old. The sensitive period is not a door that slams shut at seven weeks of age, but it is a door that slowly but steadily creaks its way to just a crack that a cat might peek their nose through as they get older. Cats who are not handled at all or much during the socialization window can still be "tamed" and learn to bond with a few people, but they might be more selective and less open to change, and it might take a lot longer for them to trust someone.

And we can't downplay the importance of the environment. A nervous cat will likely cope better in a quiet household with one or two people than they would in a home with several roommates or active children. Further, how those people in the home behave around the cat will have an additional impact; shy cats are often more comfortable with people who are relatively calm and slow moving rather than someone who is boisterous or stomps around the house!

We can make an environment more supportive to fearful cats by making sure they have lots of hiding spots and "safety zones"—boxes, tunnels, perches, or other places they feel safe that allow them to navigate through a home, from their favorite spots to their necessary resources and back again. You can also incorporate those safety zones into your play sessions!

By playing with your scaredy-cat, you are expanding your relationship with them. Instead of focusing on cuddling or handling, which might overwhelm them at times, you are giving your cat good experiences that aren't focused on handling. Although plenty of shy cats are cuddly, sometimes it takes some time and trust building to get there. Play is one great way to do so.

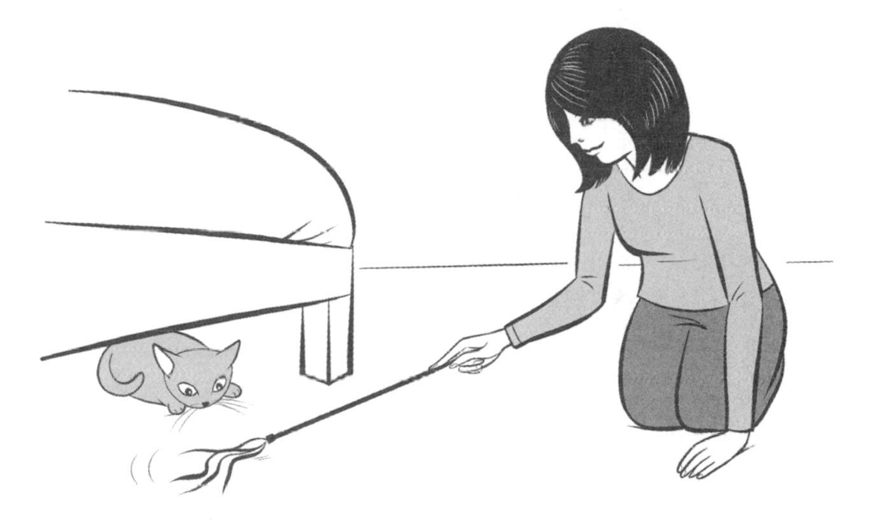

When we play with fearful cats, we have a few goals:

1. Take their mind off their fears with a positive distraction (focusing on a toy).

2. Use play to boost confidence and reduce stress.

3. Use play for desensitization and counterconditioning to help cats be more comfortable with things, people, or experiences that make them nervous (more on this soon).

My experience has been that fearful cats don't exhibit a lot of what we would call behavioral diversity. When a sound or sudden movement occurs, they have a few set responses: hiss, hide, or freeze and hope the scary thing goes away. These cats are often set in their ways, making them less adaptable to stress and change. They may have a very limited "territory" within your house, picking one or two favorite spots and ignoring the rest of the home.

THE HEALING POWER OF PLAY FOR SCAREDY-CATS

With fearful cats, we take baby steps when introducing play. By tapping into their inner hunter, we can help them engage in a very natural behavior, and break them out of the frozen state they are often stuck in. Play will diversify that body language and self-expression, and once they are chasing a toy, it can increase territory use and lure cats out of hiding. Exercise and play have stress-reducing effects, further helping your cat feel good in the moment.

- Think small—start with small toys, and slow, quiet movements of the toy.
- Use toys that look like bugs: the Cat Dancer toy, a laser pointer to get them warmed up, or use the stick end of the toy, moving it just under the edge of a towel or piece of tissue paper.
- Sit down on the floor to play so you are less threatening at first.

- Bring the toy to your nervous cat, playing within a two- to-three-foot circle around where they are.
- Never force a fearful cat out of hiding—instead give them *more* hiding and safety-zone spaces so they can expand their territory. Set up a pathway made up of boxes, tunnels, or cave beds that allow your cat to go from one safe space to the next while they play.
- Have modest goals at first—when we start working with scared cats, we hope for them to watch the toy with clear interest and little fear; eventually we hope to get them comfortable batting the toy, and before long, full-on playing on par with more confident kitties!
- A chemical assist never hurts: many cats (around 60 percent) respond positively to olfactory enrichment such as catnip. A 2017 study found that even *more* cats responded to silvervine (also known as matatabi), a Japanese plant that is more related to the kiwifruit than to catnip. Other cats love honeysuckle or valerian root. Try offering some to your cat to get them warmed up for play!
- Further better living through chemistry: if your cat's quality of life is impacted by their fear, don't be afraid to talk to your veterinarian or a veterinary behaviorist about whether medication might be appropriate.

DESENSITIZATION AND COUNTERCONDITIONING (DS/CC)

For fear and stress in response to a specific stimulus (such as a person or another cat) we rely on a process called *desensitization and counterconditioning* (DS/CC), a technique that gradually reduces fear, when done correctly. Desensitization is slow, controlled exposure—think of it like a volume knob that you turn up very slowly. You start off sitting at a distance from a fearful cat, then gradually inch closer and closer, always making sure that you aren't increasing your cat's stress.

Counterconditioning is changing an emotional response (from negative to positive) by pairing the scary stimulus (e.g., you, another cat) with something the cat really, really likes. It can be treats, it can be catnip, it could be brushing, and, yes, it can even be *playtime*!

USING PLAY FOR DS/CC

I find that a lot of fearful and shy cats can get comfortable with a human sitting at ground level but have a much harder time adjusting to people walking around (or even worse, moving quickly!). When you're small, you're safe, but once you get big(ger), you become more threatening and much harder to feel comfortable with. The goal of a DS/CC plan is to keep your cat from becoming fearful while slowly increasing your level of activity during playtime.

YOUR TURN!

Here's an example of a plan to slowly get your cat comfortable with play when they are fearful of humans. Once you have relaxation and play at step one, you can go to the next step. If you have a setback, go back a step (or two) until your cat is comfortable playing. It's important not to skip steps, and to base your progress on your cat's comfort level and interest in the toy!

1. Start by sitting on the floor or in a low chair, at a distance where your cat is comfortable. Use a relatively small and quiet interactive toy with a longer handle. Move the toy slowly, away from your cat. Is she watching the toy? If so, great! If she seems nervous or fearful, move farther back, move the toy more slowly, or call it a day and try again later, sitting a little farther away.

2. Try moving a few inches closer to your cat. If she's still comfortable playing, you're on the right track. Otherwise, go back to step one.

3. Once your cat is comfortable with the toy and showing interest in stalking it, the next step is playing while standing upright—but completely still—don't walk around just yet. Your cat may become nervous when you stand, so go back to smaller, calmer

movements of the toy at first, or move a little farther back.

4. Move the wand toy slowly toward you so your cat gets used to walking *toward* you (rather than army-crawling away) while you are standing.

5. Gradually introduce walking slowly toward and away from your cat, trying to keep her focus on the toy, not on your movements. Then start increasing the speed and intensity of your movements—always staying within your cat's comfort zone—and stop moving, or sit down as needed, if your cat seems nervous.

This process may take several days—or even weeks. Desensitization and counterconditioning is a great example of how we go slow to go fast. With time and patience, you will get to the finish line!

AVOID FLOODING

Many of us have cats who hide when the doorbell rings. One thing you never want to do if your cat is afraid of something is *force* them to get close to that scary thing (we call this forced exposure *flooding*, and it usually increases an animal's fear). People often have good intentions and want to "show" their cat that a visitor is safe by pulling them out from under the bed (never mind that the cat is clinging to the carpet for dear life), and even placing them on the visitor's lap. "See, they are perfectly nice!"

Your cat is left feeling even more afraid and less in control of their fate. They told you that they were afraid by choosing a place to hide, and you had to go and ruin their safety. Their fear of the doorbell has been solidly reinforced.

Okay, enough admonishment. What's a good alternative in this scenario? Well, with shy cats, sometimes the outcome is just that they are allowed to stay in the safe space of their choosing when there are visitors. But in some cases, you might be able to tempt your cat to show their face for a little playtime—if the visitors are sitting quietly! Again, return to the basic principles: the volume knob (controlled, slow exposure), pairing positives (treats, play), and avoiding flooding. Because your cat will be more afraid with strangers in the home, start by sitting on the ground using slow movements with a quiet toy. You may be pleasantly surprised how your cat will make a guest appearance when they are tempted by a toy and given the choice to stay hidden or retreat as they need to.

NIGHTTIME ACTIVITY

"He won't let me sleep" is one of the most common complaints I receive from clients. I've talked to many humans who were literally in tears from the sleep deprivation caused by their cats pouncing on them at 4 a.m. or begging for food throughout the night. People mistakenly refer to cats as nocturnal, but remember that cats are most active at dawn and dusk. Their activity cycles are also highly influenced by our presence and activity.

We tackle this problem from multiple angles:

- Keeping the cat more active during the day
- Giving them a routine that helps them settle and sleep through the night
- Ignoring attempts to get you out of bed for attention or food during the night or early in the morning

In the case of the night owl, more play will *never* hurt. In fact, as a test, I often have my clients play as frequently with the offending cat as they can each day for several days in a row. Yep, just keep playing. No, you're not done, play some more! Your cat is napping? Get some jingly feather wands with bells and start wiggling them until he's tempted by the sound. The cat guardian often gets a few days' respite and a few nights' good sleep. Of course, that amount of play isn't necessarily sustainable, but it tells us that with enough physical and mental stimulation during the daytime, it is possible for them to sleep through the night.

- Increase play sessions to three or more times per day. They do not necessarily have to all be long, but they should have some level of intensity that helps tire the cat out. The sessions are ideally spaced out through the day. And make sure to end each play session with a treat.

- The most intense play session of the day should be reserved for close to your bedtime; when the play session is over, give the cats their largest meal of the day. I've noticed that some cats seem to get a small burst of energy after a meal, even when you played with them before eating, in which case you may want to allow for about twenty minutes between your cat's last meal of the day and your bedtime.

- Ignore any feline activity during the night; if you get up and feed the cat or give them attention when they pester you, you are rewarding the cat and the behavior will continue!

- Do not feed the cat first thing in the morning; if so, they will try to wake you up early, anticipating that their breakfast will also come earlier. Instead, get up, have your coffee, take a shower, and *then* feed your cat, preferably at a consistent time each morning. It should be the time of day, not you getting out of bed, that predicts the arrival of food in the dish.

- During the day, provide the cat with more enrichment and activity. This can be in the form of food puzzles, novel items to explore, clicker training, harness or stroller walks, olfactory enrichment such as catnip and silvervine, and automated toys (see chapter 13).

- There are different ways to help your cat settle at night. Some cats respond well to a heated pad that is only turned on at night—irresistible! You can also talk to your veterinarian about supplements, calming diets, or even medications that can help you and your cat get those precious *z*'s.

- In some cases, I do recommend keeping kittens or young adult cats out of the bedroom, especially when they are new to the household. They can either be confined to a nice, cozy room with everything they need (water, food, litter box, bedding, safe toys, scratching post) for a few weeks until they are in the routine of settling down to sleep at night, or you can close your bedroom door at night. Eventually many of these cats can be transitioned to sleeping in your bed with you (or at least having the freedom to if they wish)—the key is just that they first learn that nighttime is quiet time, not attention-, play-, or getting-fed-time.

CAT, MEET THE OTHER CAT

Introducing cats to each other is a process that has around a 9 percent failure rate, according to the scant research on cat relationships. I wish there was a magic wand that would allow cats to say, "Hey, let's be friends." On the flip side, about two-thirds of cats will grow to tolerate a newcomer safely within a month, when introduced slowly and in a controlled manner (in other words, not letting them "just fight it out"). Play is just one piece of the puzzle in helping cats accept a newcomer into their territory, but it can be a very important piece, serving multiple functions.

At this point, I hope you feel comfortable with the concept of DS/CC. We use it for introducing cats to each other too! The first steps of a cat introduction include keeping them separated for a few days, letting them get acquainted with each other's scent and sounds with the safety of a closed door between them. Next, we start visual introductions between cats with them at a distance from each other, ideally with a baby gate or screen between them. We would gradually (and slowly) move them closer to the screen. That's the desensitization part. For counterconditioning, we want the cats to be having positive experiences with each other. In this case, it's whatever each individual cat likes—it could be treats, trick training, catnip, brushing, or *play*! I like to use play for this process because it gets the cats showing relaxed, nonthreatening body language around each other. Cats don't have a lot of facial muscles to express subtle emotions, and they don't have many appeasement gestures (behaviors that say "I'm friendly and mean no harm"). Instead, they rely heavily on larger signals and changes in body language to inform them of another cat's intent.

Cats are also social learners and depend on cues from other individuals (humans or cats) to tell them if the environment is safe. If you are a cat, and another cat is in the same general space, seems to be having fun, and isn't behaving as if they are scared or threatened, you can assume that it's also safe for you to play!

What is positive for one cat might not be for the other (one cat might prefer playing, while the other instead gets treats), if one cat plays very enthusiastically, you must make sure that their zest for toys isn't threatening or off-putting to the other cat. In that case, switch to more "ground play" rather than encouraging leaping, or try a smaller, quieter toy.

Play helps relieve tension between cats because it's a positive experience, but also because it's incompatible with undesirable behaviors. A cat cannot stare at, stalk, or chase another cat if they are focused on a toy! In this case, you are guiding your cat toward things you want them to do, making it harder for them to do things you do not want them to do.

To that end, sometimes we have an energy mismatch, where one cat is much more playful than the other. Much like the play-aggressive cat may stalk your toes, they may also ambush and torment other cats in the house. As discussed in the chapter about social play (chapter 8), we must take extra effort to make sure that the more playful cat gets enough exercise to tire them out. Many of the same techniques for dealing with play aggression toward humans can be used to distract cats away from other cats—tossing toys at the perfect moment to prevent a pounce, using toys to direct a cat away from other feline members of the household, and generally increasing the amount of playtime the more active cat receives.

We can also use toys to boost the confidence of "victim" cats, much like we do with fearful cats. Cats who chase tend to feed off the energy of the other cat; a cat who runs will be chased. Just as we want to get the rambunctious cat less motivated to chase, we also want to make the victim cat less likely to be chased in the first place. To keep things safe, this often means playing with the cats with a baby gate between them, or even playing with them separately (that means different rooms with the doors closed)—I'll go into this more in the next chapter, when I address playing in multi-cat households. The benefits of play are more than just transitory—when the cats meet again, they will both be more tired and relaxed, hopefully more inclined to nap than argue.

BABY GATE

WHEN YOUR CAT IS PEEING OUTSIDE THE LITTER BOX

I think nothing inspires dread in shelter workers, behavior professionals, or cat owners like a puddle of cat urine on your floor, on your bed, on the sofa, in your shoe, in the fruit bowl, or on the kitchen stove (I've seen clients with cat pee happening pretty much anywhere you can imagine). Unfortunately, many cat owners immediately focus on the property damage, and not on their cat's cry for help. House soiling primarily occurs for one (or more) of these three reasons: a medical problem, something intolerable about the litter box, and stress. Rather than investigating and working to solve the problem, many owners have a no-tolerance policy: litter box avoidance can send cats to the shelter quickly, and even to their death (euthanasia for behavior happens more than we like to talk about).

The most common cause of lower urinary tract signs (such as blood in the urine, urination outside the litter box, or straining in the litter box) is FIC, which is short for "feline idiopathic cystitis." FIC is a medical condition with an "unknown" cause, the definition of *idiopathic* being "any disease or condition which arises spontaneously or for which the cause is unknown." Without a known medical cause, FIC is considered a diagnosis of exclusion, where tests are done to rule out other conditions (such as urinary tract infections, urinary crystals, or kidney disease).

If your cat isn't using their litter box(es) consistently, the first thing to do is take them to the veterinarian for a full exam, including a urine test. Cats with FIC often need short-term medications to help with the pain and inflammation they are experiencing. The second thing you need to do is take a long, hard look at your litter box setup and make sure it suits your cat's needs (not just yours): Do you have multiple boxes (at least one per cat, ideally plus an extra)? Are they clean enough, big enough, easy enough to access, with a litter that your cat prefers (usually a fine, unscented, clumping litter)? There are plenty of resources out there on setting up the ideal litter box, and my advice: Keep it clean and keep it simple. Avoid robots, pellets, and self-cleaning anything and you're off to a good start.

Now you must address the third piece of the pee puzzle: Is your cat experiencing stress?

What is stressful to your cat might seem trivial to you. But research by Dr. Judi Stella and Dr. Tony Buffington showed that changes as simple as what time cats were fed and who was feeding them could cause cats to display "sickness behaviors"—loss of appetite, vomiting, diarrhea, reduced activity, hiding, skin problems, and urination or defecation outside the litter box. I hate to paint cats as delicate snowflakes, but this tells us that consistency and predictability *matter* to them.

HELPING CATS WITH FIC

The most effective treatment for FIC is modifying the cat's environment to increase their perception of control and decrease their feelings of threat (in other words, we are trying to reduce stress!). We can do this by using a model developed by a group of feline experts to help ensure that cats' environments are set up for success. "The Five Pillars of a Healthy Feline Environment" include safe spaces, multiple and separated resources, a respect for the cat's sensitive nose, positive social interactions, and my favorite pillar: opportunities for play and predatory behavior!

Playtime can help reduce the effects of FIC. You've probably heard that exercise is good for you—like in humans, it is believed that other animals also experience an exercise-related endorphin boost that can improve mood and reduce feelings of pain, stress, or depression. Both aerobic (moderate) and anaerobic (high-intensity) exercise have been associated with increases in beta-endorphins, which are part of the endogenous opioid system of the brain (one way that the body regulates pain and experiences reward).

For cats with FIC, interactive play should be tailored to the needs of the cat, considering their age, personality, and play preferences. Because routine is so important to these sensitive kitties, schedule your play sessions for around the same times each day. Although there's really no "special way" to play with these cats, my take-home here is that playtime, however best suits your cat, is an important part of the treatment.

And there are *plenty* of reasons to be hopeful about cats with FIC. Research has shown that environmental enrichment can be very effective in treating cats with litter box avoidance, with around 75 percent of cats in one study responding positively to the changes their humans made, and experiencing improvement in signs of FIC.

IT'S NOT MAGIC, BUT IT CAN BE MAGICAL

Much like play alone isn't going to guarantee that an overweight cat can get to a healthy weight, play alone cannot "cure" most behavior problems. Behavior does not happen in a vacuum, so when we are working to change behavior, we must determine why the behavior problem arose in the first place. Play can fit into a behavior modification plan for many reasons—to increase exercise and reduce stress, to provide an outlet for predatory behaviors, to build confidence, to prevent boredom, or to enhance the bond between a cat and human. To see these changes as a result of playtime is one of the most satisfying parts of the work I do—whether it's seeing a scared cat go completely nuts over a feather wand, or a normally rambunctious cat exhausted and purring contentedly next to their person. Although results may vary depending on the cat, in every case, we are making moves toward greater feline happiness and greater owner satisfaction.

CHAPTER 12

Playing with Cats When They Don't Fit the Mold

Giving advice is a tricky thing. Even when people pay me good money for cat behavior advice, it is often met with resistance: "Well, what about (insert excuse X, Y, and Z)?" "That won't work for my cat because . . ." So many cat owners are looking for any reason to not change their *own* behavior, in the hopes that their cat will just transform into a perfect pussycat without any human effort. If you throw up your hands and put up obstacles, it's easy to just give up rather than pick up that wand toy. But I truly believe that play for cats is a right, not a luxury, and we humans must be ready to adapt.

I've already given lots of advice on how to play with your cats—understanding what your cat brings to the table, how to choose toys, how to move the toys, and how to use play to prevent and fix behavior problems. But some of those recommendations may not fit if your cat has special needs, for example, if they're blind, or if you've got two . . . or three . . . or more cats. The focus of this chapter is to address the special circumstances that might apply to your cats, or cats you will encounter in your lifetime. In general, it's not a radical shift but just a slight adjustment in your technique and approach. While I hope I've given you the tools to build intuition and fine-tune your methods to match any kitty you happen to be playing with, this chapter will give you some further ideas for playing with the cats who break the mold.

OLDER CATS

Many folks assume that older cats don't play, because their play doesn't always look like it did when the cat was younger. There may be no backflips or frenzied parkour jumps; interest in a toy seems to end just as it begins; and you start to wonder: If this is a simulation of hunting behavior, could a twelve-year-old cat even survive in the wild?

First, a disclaimer: Some cats do stay pretty darn frisky throughout their lives, and that's a wonderful thing. But it's the cats who visibly slow down, leading their humans to stop trying to play with them, that I'm most concerned about. Multiple studies in mammals (not just humans) show the benefits of various types of physical and cognitive exercise in preventing or delaying some of the changes that come along with aging. For humans, staying active staves off muscle wasting, cognitive decline, and physical frailty, and increases life expectancy. There's no reason to think that cats don't enjoy some of these benefits as well, and I've seen an increased interest in research on the aging feline, so perhaps soon we will know for sure.

Cats age much faster than we do and are experts at hiding pain. The most important thing you can do for your senior cat is to take them for twice-yearly checkups. This allows you to stay on top of any brewing health problems and to address them so that your cat feels good and enjoys the greatest quality of life for the longest period possible.

Next, adjust your expectations: Your senior cat may just bat at the toy or even watch it, and that's just fine at first. As you find your cat's favorite toys and movements of the toys, you may find that their inner kitten awakens. The backflips may follow.

TIPS FOR PLAYING WITH SENIOR CATS

- Focus on slow movements: you might have a weaker predator, so make sure the prey is weaker too.
- Try moving the toy under towels, blankets, tissue paper, or between couch cushions. The Ripple Rug is a great tool for creating an irresistibly elusive playground of toy hiding places.
- Try small, quiet toys to pique your cat's interest at first: think more bugs than birds.
- Use ramps and soft surfaces to give your cat support. Ramps will allow your cat to easily access vertical spaces.

Carpets provide traction that your cat's claws can dig into, and yoga mats and rugs provide a little padding in case of slips or falls.

- Take advantage of your cat's naturally active periods: even senior cats get the zoomies. By knowing your cat's internal rhythms, you can make it more likely they will respond to your play maneuvers.

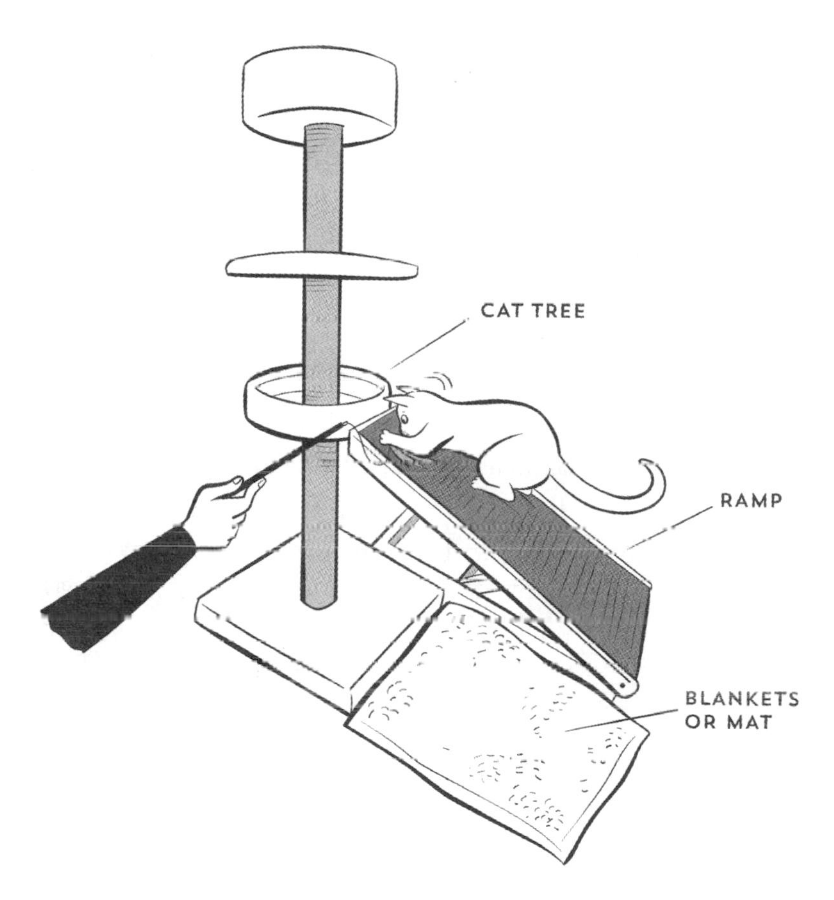

OBESE AND OVERWEIGHT CATS

Much like aging cats, obese or overweight cats are victims of the "they don't seem interested, so why even try?" attitude. We also forget that, for overweight cats, that extra body weight can be *painful*. In the short-term, it makes movements less fluid and more difficult, and in the long-term, that extra weight they're carrying around can lead to mobility problems and joint degeneration. This can make the active parts of play a bit uncomfortable until some of the ounces melt away.

If your cat is overweight or obese, I strongly encourage you to get your cat started on a weight loss program, but you *must* work with your veterinarian to do this safely and effectively. Cats who lose weight too quickly are at risk for developing a condition called *fatty liver disease*, where the liver becomes overwhelmed by the breakdown of fats. Fatty liver disease can quickly become fatal. For this reason, it's important to have a veterinarian-approved plan for slow, gradual weight loss (usually through dietary changes, increased activity, and frequent weighing of your cat to track progress).

Keep in mind that play with an overweight cat might appear a bit awkward at first. They really might not do much but roll around and watch the toy. That's okay right now. The most important thing is to meet your cat where they are, and to gently ramp up the play as your cat becomes more fit.

TIPS FOR PLAYING WITH CHUNKY CATS

- First, see the section on senior cats above, as a lot of the same rules apply!
- Focus on ground play at first: move toys close to the ground so your cat can bat at them.
- Use ramps and encourage climbing. If you can get your cat to step or jump up onto low furniture (like a coffee table or step stool), that's a great start. If your home has stairs, try slowly dragging a toy up and down the staircase (preferably with carpet or antiskid mats for secure footing).
- Consider harness training your cat for some safe outside walks, which will provide both mental stimulation and exercise. You can even play outdoors if you bring a wand toy along.

I hope that now you and your cat feel ready for the "couch to Cat Dancer" program. It's a slow but steady race, and the finish line is a healthier and happier cat!

YOUR TURN!

If there's one thing most overweight cats are, it is food motivated. Why not use that to your advantage? One of my all-time favorite games for chubby cats who love food is what I call "kibble toss." It's just as it sounds: you skitter one piece of dry food (or a treat) at a time across the floor. I've never seen some cats run so fast! To take full advantage, toss the kibble down some stairs, or get your cat to come back to you (we call this a "recall") in between tosses, or toss the next kibble in a different direction. Otherwise, your cat may just stay in place to anticipate your next toss, maximizing food gain while minimizing effort (we are trying to reverse this trend, but hey, there's a reason your cat became overweight in the first place). Don't forget: those tossed treats *count* toward your cat's daily rations, so be sure to subtract them from the calories you will offer at mealtime.

CATS WITH DISABILITIES

It's no longer unusual for folks to rescue cats who may be blind, deaf, or have issues with mobility. In the past, these cats may have been seen as less adoptable, but now we recognize that they are not only adoptable but also highly adaptable! Caring for a special-needs cat can be a deeply bonding experience; they have plenty of love to give and can enjoy life to the fullest with just a few adjustments to the environment and how you interact with them. And like all other cats, these cats need and want to play!

1. BLIND OR VISUALLY IMPAIRED CATS

Some cats are born blind; others may lose vision or even an eye or two due to infection, injury, or disease. A cat who has been without vision their entire life, or who gradually loses their vision, will typically adjust seamlessly compared to a cat who loses their sight suddenly. Regardless, blind cats quickly adapt, using their whiskers, hearing, and sense of smell to find their way around your home.

There's nothing like meeting a blind cat to drive the point home: cats do not rely on vision the way we do! When I went to one client's home, their cat, Neko, jumped from the floor up onto the kitchen table to greet me without missing a beat. He had no eyes, so I knew he was using his whiskers, ears, nose, and muscle memory to find me to say hello. And yes, he loved to play!

TIPS FOR PLAYING WITH BLIND CATS

- Keep the play environment consistent: while you can introduce some novelty in the form of props, you want to make sure the main stage (e.g., the human and cat furniture and any elevated surfaces) does not change suddenly.
- In general, avoid picking up and carrying blind cats. They rely heavily on their cognitive map they have

created with their nonvisual senses. Bring the playtime to them or call them to play with your voice or the sound of a toy.

- Don't startle your blind cat—especially if they are sleeping. Talk to them when you want to start the play and let them orient to the sound of you and the toy.

- Blind cats can play with almost any cat toy except a laser pointer! When you are choosing toys, think about the kinds of sounds they make when you move them. Your cat may have sound preferences (e.g., bells, crinkle toys). Moving the toy a foot or two away from your cat will create enough air movement that their whiskers can home in on the toy's location. And don't forget to let your cat enjoy plenty of tactile experiences with the toys—let them chew and grab to their heart's desire.

2. DEAF CATS

Like blind cats, deaf cats can be born deaf or lose their hearing due to injury or illness. Some people don't even notice their cats are deaf until long after the fact—after all, cats are already known for their "selective hearing" (this is especially true when cats become deaf gradually, as they often adapt just fine to the slow hearing loss). People often use a flashlight or hand signals to communicate with their cats, much like we would do when we call our cat's name. Your cat can quickly learn that a hand signal means treats are available or it's time to play. Playing with a deaf cat just requires a few minor accommodations.

TIPS FOR PLAYING WITH DEAF CATS

- Avoid surprising your deaf cat, especially when they are at rest. Since they can't hear your approach, make sure they can *feel* it. Tap the floor or furniture where they are, rather than touching them unexpectedly.

- Move toys slowly into their field of vision—ideally along a horizontal plane at first, which is what their eyes will be most sensitive to. Once they have sighted the toy, you're good to go!

- Although we can't use those rustling sounds to entice a deaf cat, you can tap the floor or a piece of furniture with the toy. Your deaf cat will sense the vibrations and hunt accordingly.

- Deaf cats may enjoy loud toys that typically scare other cats, such as robotic toys. However, it's not just the sound that can scare cats with hearing, so aim for smaller toys with small, lifelike movements.

- Think lights and sparkles—things that will catch your cat's attention! Many Mylar toys appeal due to their sound, but they also glint when they are hit by light, which can make them attractive to deaf cats.

- Don't forget textures and scents. Your deaf cat's other senses are likely heightened, so tap into that opportunity to let them taste, touch, and smell their toys.

3. MOBILITY ISSUES

Cats can have mobility issues for different reasons—from experiencing the removal of a limb to having a chronic neurological condition. But if you've noticed a trend here, just like cats with other disabilities, these kitties still love life, and they still love to play!

Most three-legged cats adjust surprisingly quickly to the loss of a limb—it's much better for them than having a painful leg. Soon you will forget that your cat had a leg removed, as your tripod zips around the house as if they had an invisible fourth limb. For cats who lost a hind leg, they may need help with climbing and jumping *up*; and for cats with a missing forelimb, the landing may be more challenging. But they will still jump—perhaps just not as high as they used to (but don't be surprised if they do!).

Paralysis in cats can range from weakness in the limbs and body to complete inability to voluntarily move different parts of the body. The causes range from disease to chemical exposure to trauma. Some causes of paralysis are quite serious and often result in humane euthanasia for a cat because their quality of life will be poor. But many cats with partial paralysis can experience a life full of joy with some adjustments and diligent care (such as a wheelchair or cart, booties or wraps to protect dragging limbs, and diapers).

Finally, cats with a disorder called *cerebellar hypoplasia* (CH) have a neurological condition where the cerebellum (the part of the brain that controls motor activity) does not develop properly.

Kittens are impacted in the womb if their mother is infected by the panleukopenia virus. The severity of their CH can depend on when the mother was infected and the level of cerebellar development at the time of infection.

CH kitties typically have uncoordinated movements, such as goose-stepping, teetering, and sometimes falling over. Most kittens with CH get along just fine, and their condition is not painful or contagious. Aside from needing some protection from injuries that could come from a serious fall, cats with cerebellar hypoplasia don't seem to know anything is the slightest bit amiss, and they are perfectly happy to get along with their day (including plenty of playtime), thank you very much! (And if they just fell over, well . . . they meant to do that!)

TIPS FOR PLAYING WITH CATS WITH MOBILITY ISSUES

- Start with gentle exercise and work up to upright play, vertical surfaces, or climbing based on your cat's abilities.
- Play may be low to the ground for these kitties, so focus on ground movements (think mice and crawling bugs).
- Tripods and CH kitties appreciate some padding around the play area to protect them from slips and falls. Cushions and nonslip rugs can provide further opportunities to grip and support themselves.
- With cats who can pull themselves around with their front legs but don't have mobility in the back legs,

sometimes a combination of carpet (for grip) and hardwood or tile (for slip) allows them to move around for play without too much friction.

- Work their strengths: all of these cats can still enjoy gentle play and stalking, and they can touch, smell, and taste toys.

4. DECLAWED CATS

I wish I didn't have to write this section of the book, because declawing is a cruel procedure that causes a great deal of suffering in cats. Declawing is the amputation of the last bone of each of a cat's toes (imagine suddenly having the tips of all your fingers removed), typically performed to stop cats from scratching furniture and humans, or preventively as a package deal with spay or neuter surgery. Essentially, it is an elective surgery that leaves cats disabled. Declawing is illegal in many countries and in some US cities and states, but it is still frequently performed (although there are no exact figures, about 70 percent of US veterinary clinics offer the procedure, and it is estimated that at least 25 percent of owned cats in the United States have been declawed).

Let's be clear: this surgery is done for human convenience, often because cat owners do not provide their cats with appropriately tall and sturdy scratching posts and, in the case of kittens who are declawed when they are spayed or neutered, before they are even given the chance to be trained to use a scratching post. Declawing offers no benefits to a cat, and it can cause many harms.

Cats naturally walk on their toes, and when their toes are amputated, they must adjust to walking in a flat-footed fashion, on their wrists. Changing their posture can lead to joint problems later in life. In fact, in a study of 274 cats, it was found that declawed cats were more likely to experience pain and behavior problems than cats with natural claws. All cats received a veterinary exam, and the 137 declawed cats also had their paws x-rayed. Around two thirds of the declawed cats in this study had bone fragments remaining in their paws, likely causing an uncomfortable "pebble in the shoe" sensation when they walked or used their paws.

You can imagine that such pain is not conducive to play. A 2017 study found that owners of declawed cats reported them as less playful and less predatory than their clawed counterparts. The authors of the study interpreted this finding as suggesting a negative impact of declawing on the welfare of cats.

If your cat has been declawed, they can and will still play. But you should approach play armed with the knowledge of how their physiology has been altered, and how pain and a lack of claws could impact their general wellness. A cat in pain is less likely to play, but when pain is controlled, your cat can enjoy most aspects of play!

TIPS FOR PLAYING WITH DECLAWED CATS

- Please get your cat's paws x-rayed for fragments. Show your veterinarian videos of your cat walking and playing to help them assess your cat's mobility and comfort. Many declawed cats benefit from pain control and repair surgery, which you can discuss with your veterinarian.

- Cats grip with their claws; if your declawed cat still has their back claws, they will use them when they are playing, especially with longer lures or kick bags they can attack with those back feet!

- Depending on your cat, they may enjoy play lying down on their side or on their back to reduce pressure on their feet.

- Adjust play so that your cat has more chance of success without having to hook the toy in their claws. Move the toy slowly, allow them to grasp toys with both paws, and give them time to use their mouth and whiskers to investigate.

- Use toys with textures your cat enjoys biting, such as cardboard, kraft paper, or Mylar film.

- Don't forget to appeal to your cat's sense of smell and sound.

- Declawed cats still have all their instincts and often go through the motions of scratching, grasping, and bunny-kicking. They need outlets for all these behaviors, such as kicker toys (yes, I even recommend scratching posts for declawed cats).

PLAY IN THE MULTI-CAT HOUSEHOLD

Playing with multiple cats is like taking everything I've shared in this book and multiplying it by two (or three or four or . . . ?). So much about managing play in the multi-cat household is about understanding each individual cat's preferences and needs for play, and understanding the relationships between your cats (do they get along? Does one intimidate the other?). Now you try your darn best to meet all your cats' needs while still having time to make a living and sleep. You might be wondering: If you have three kittens, and I said each one might need up to two hours a day of play, are you on the hook for six hours of play a day?

Absolutely not! You made the smart choice in adopting multiple kittens who hopefully enjoy romping with one another. That doesn't take away all your responsibility, but it can help. On the other hand, I don't recommend having more cats than you can provide for—and we're not just talking about space, food, and medical care; attention and playtime need to be factored into that equation. If you have three adult cats who don't enjoy playing with one another, and all have different play interests, you are going to have to put in three times the time and effort to make sure they all get some play, every day.

The biggest mistake I see cat owners make is assuming that some of their cats aren't playful. "He just likes to watch the other cat play" is code for "One cat hogs the toy, and the other cat is too intimidated to get in there and romp"! We often don't notice feline intimidation because it can be quite subtle (you may not see fighting; cats often choose avoidance instead of physical conflict), but when the toy hog is secured away in another room, you'll often notice that the cat who "just likes to watch" seems more than happy to jump into the fray.

TIPS FOR PLAYING IN MULTI-CAT HOUSEHOLDS

- Know your cats! If you have a toy hog or a "just watching" cat, you need to adjust playtime so that everyone gets a fair shake at fun.
- Know how well your cats get along. If they accidentally bump into one another during a shared play session, is it likely to start WWIII? Or will they brush it off or even turn it into a little good-natured wrestle?
- Your cats may have different activity cycles, play styles, and toy preferences, which should all be factored into your play plan.
- Play with the cats separately if necessary. Yes, this might mean you have to secure one (or more) of your cats in another part of your home. Baby gates, screen doors, and regular doors can all be your friend. However, this isn't meant to be a punishment for the secured cat—it's a good opportunity to provide them with an exciting solo-play toy or food puzzle (see the next chapter) to keep them busy.
- If there are multiple humans in the home, split up the play duties accordingly. Two people can entertain twice as many cats as one person can.

- A multi-cat household is a very good reason to have lots of different toys and activities, so that every cat has a chance to engage with something they enjoy during a session.
- Get good at moving two toys at once (yes, try having a wand in each hand! It's kind of like patting your head while rubbing your belly).

HOW I MANAGE PLAYTIME WITH THREE CATS

One way to manage play with multiple cats is to sneak in play with each one when the others are napping or otherwise engaged. For example, if one of my cats comes in from the catio, but the other two are still outside, that's a great opportunity for a short one-on-one sesh. However, my three cats are also sisters with terrible FOMO, so I must be prepared for another cat to show up expecting some playtime too. But sneaking in a few minutes here and there works well at getting everyone some quality solo toy time.

For our formal "family meeting play sessions," I have a slight advantage in that despite having three cats, we also have two humans. Despite having four arms between us, my boyfriend and I usually have only one toy each. We first set up multiple play areas in different parts of the house. One cat might be rolling in some silvervine with a kicker toy in the bedroom, while one is playing with the feather wand on the couch, and another cat is on the Ripple Rug chasing the mouse on a stick. Then there is some swapping of roles. Sometimes all three cats want to play in the same space, and other times they naturally separate themselves. There is a lot of fluidity and flexibility, and some nights it's better than others. But for the most part, we make it work!

SEPARATE PLAY "STATIONS"

SHELTER CATS

Some of you may work in animal shelters, or I hope at some point in your life you consider volunteering in an animal shelter! It's so rewarding and leads to so much valuable experience with cats, and the time you spend with shelter cats is critical for them as well. Shelter cats are just cats, but they are cats in a very difficult situation. Shelters can be incredibly stressful places for cats; cats are very much creatures of habit, and now they are in a loud building filled with strange people and other stressed animals. All their senses are continually inundated with sounds and smells and handling that might not be pleasant. Shelters often rely on volunteers to give cats positive interactions—whether it's just reading to them quietly or petting them or, if they are open to it—playing with them!

Unfortunately, many cats in shelters are too stressed to play, especially at first. Many cats will start feeling less stressed after a few days, and that's a good time to learn more about them and see what kind of interactions they enjoy. So much will depend on the cat's age, personality, and health, but also on the shelter: Do cats have rooms to stay in or cages? Are they housed with other, unfamiliar cats? Is the shelter set up to offer cats hiding spaces? Are the staff, volunteers, and visitors loud or quiet? Are people given training or direction in how to safely interact with cats in the shelter?

A 2017 study gave twenty-six shelter cats a simultaneous choice between four chambers—one with a hiding box, one with toys, one with a perch, and one empty control chamber. The food, water, and litter were all in the central area, with the four chambers attached to form an *X* shape. The cats spent around half of their time in the hiding box, regardless of age, sex, and whether they were a stray or owner-surrendered. They spent very little time in the chamber with a toy.

Even though it might seem like giving cats a box to hide in might hinder their chances of adoption, a 2014 study found that when cats had a hiding box, they adjusted to being in the shelter faster. When they didn't have a hiding box, they tended to hide behind or in their litter box. These findings tell us that hiding is an intrinsic need of cats, especially when stressed, and providing cats with a box in a shelter or similar environment should be considered a necessity. And hopefully once those cats are adopted into homes, boxes become part of playtime and not just a place to hide!

Although some cats will be too scared to play in the shelter, others will be more than ready and will appreciate the opportunity! The young, outgoing cats may have a lot of pent-up energy and be a bit desperate for play (these cats are often quick to nibble at approaching hands, and their cage may be a mess from them throwing litter and cat food everywhere!). As for the shy cats, once they are ready, playtime can help them forget the stressors of shelter life and give them a moment of confidence.

When playing with shelter cats, there are always going to be constraints. Toys must be disinfected between cats to prevent spreading germs, and certain toys may not be allowed for safety reasons. You are going to be working in a much tighter space than you would be with cats at home, and there will be many more distractions, including other cats meowing, humans visiting, dogs barking, and cage doors slamming. Be patient, do your best, and use the tools you've gathered to give shelter cats a chance to play.

TIPS FOR PLAYING WITH SHELTER CATS

- Always check with shelter staff, as shelters have protocols and rules that take precedence over my suggestions! This includes rules about sanitizing toys, which toys are allowed, and any safety protocols to keep you and the shelter cats safe.

- I've said it before, but I'll say it again: Please don't use your hands as toys. Cats who play rough may have challenges getting adopted, and if shelter cats bite or scratch humans, they may have to be quarantined, depending on shelter protocol (and often state laws related to rabies exposure). Do everything you can to help shelter cats direct their energy toward toys, not hands.

- Use small toys, especially for scared cats.

- Watching a toy—with interest and without signs of fear—is positive, but watch for signs of fear (making their body smaller, backing away, cringing). If a cat seems afraid, they might not be ready for play, or the toy and its movements might be too scary.

- Move toys slowly, and if there is bedding in the cat's cage or room, move the toy under the edge of a bed, towel, or blanket for the good ol' mouse-under-the-rug trick that most cats love.

- Be sure to wind down your play session by moving the toy slowly before you stop your visit, so the next visitor isn't taken by surprise by a cat who is ready to pounce!

- Make space to play: if allowed, move the litter box, food, and water out of the cage during your visit to make more room to move the toy. Don't take away a cat's hiding spot.

- If you are playing with a very active and confident cat, and your shelter has a visitors' space, see if it can be used for play sessions!

- Find out what the shelter's policy is on enrichment for cats. Ask if all cats can have safe solo-play toys (such as

Ping-Pong balls) left in their cage—research shows that many cats are more likely to use toys and other enrichment overnight, when the humans aren't around!

CATS WHO GO OUTSIDE

This is perhaps the most controversial aspect of keeping cats as pets: Should we keep them indoors or let them roam freely? There are strong feelings on both sides, for a multitude of reasons. In the "keep them indoors" corner, we've got cars, fights with other animals, and the potential for your cat to become a mass killing machine once they're unleashed at the neighbor's bird feeder. In the "free, free, set them free" corner is that cats get more exercise and can express more natural behaviors outdoors than in.

The whole premise of this book has been providing a way to give cats the key to one of their most basic instincts—hunting. And yes, the inspiration for this book was in large part to give indoor cats a better outlook on life through more exercise and enrichment. Indoor cats are prone to obesity and stress. However, that doesn't mean that only indoor cats will benefit from playtime. And it also doesn't mean that the grass is always greener on the other side of the cat flap. Outdoor cats do have their fair share of stressors, including territorial battles with neighborhood cats. And quite frankly, the statistics for obesity and stress-related health problems aren't stellar in areas where cats have more frequent access to the outdoors.

Now, I'm not going to come to your home and make you seal up your cat flap, and I certainly won't deny that cats enjoy their outside time (I would also argue that there are many other safe ways to give cats fresh air, including catios, pet strollers, and harness training). But I would like you to consider whether you're using "my cat goes outdoors" as an excuse to not provide your cat with many of their basic needs in the comfort of their home. My experience has been that cat owners often see the cat flap to freedom as justification to not give their cat (among other things) a litter box, indoor climbing areas, or playtime.

Why does your cat need play if they go outdoors? One of the key reasons to play with your cat is to build your bond and strengthen your relationship with them. Even if your cat wanders—whether just a few yards away or a few blocks away—your home should still be their most important territory. You can do that by providing them with all their needs indoors, and by making play a part of your daily routine.

Also, have you ever watched outdoor cats? They spend an awful lot of time sitting around lying in the sun. Not exercising. I don't know of any research to date that has actually compared the activity levels of indoor cats with those who go outside, so I'm skeptical that they are engaging in the feline equivalent of a triathlon. For now, the jury is out, and you can assume that a little extra exercise indoors will not hurt your indoor-outdoor cat.

A 2021 study did find some differences between indoor cats and those with outdoor access when it came to play. Indoor-only cats were quicker to chase a ball than the indoor-outdoor cats, but the difference between the groups was small. There were no differences between the cats in their likelihood to go for a feather wand. Let's just say, it's time to put your excuses away and dust off that pile of interactive toys.

TIPS FOR PLAYING WITH CATS WHO HAVE OUTDOOR ACCESS

- I recommend establishing a schedule for play (as for other cats); before mealtimes is always a great idea.
- Is your cat a hunter? If so, they have seen and killed the real thing, unlike cats who have been raised indoors. Think about what prey they tend to bring home for you—are they a generalist (hunt anything that moves) or a specialist (strongly prefer one or two prey types)?
- You may need to try uber-realistic toys: look for toys with real fur and feathers, or at least acceptable substitutes.
- Play with your cat where they are: bring some toys outside with you and try playing with your cat outdoors.
- Conversely, bring the textures and smells of the outdoors inside—your cat may enjoy playing with leaves and branches, either as toys or as an "enhancement" to your play area.

CHAPTER 13

Other Ways to Inspire Your Cat's Inner Hunter

Although the overarching theme of this book has been that the best way to connect with your cat and simulate the hunting experience is through interactive play (and I stand firmly behind that assertion!), there are other activities and objects that can provide our cats with predator-appropriate fun.

What does it mean to tap into our cat's natural lifestyle as a hunter? Let's consider what they do naturally when they hunt. We've already given cats plenty of opportunity to stalk and rush by implementing interactive play. But that's not all that cats do when they hunt. Cats also:

- Work for their food
- Use all their senses (including scent and touch)
- Explore their environment

Before we conclude, let's dive into a few more ways to re-create the hunting experience(s) for our cats, considering a diverse buffet of activities, settings, and behaviors.

WILL WORK FOR FOOD

As I have drilled into your head throughout this book, cats are natural hunters. It follows that, in the wild, they must put effort into acquiring food. But we've drastically changed this way of life for our cats. We brought them inside, threw their food in a bowl, and essentially took their jobs away. Now unemployed, our cats may become bored or frustrated, or engage in undesirable behaviors. Let's put them back to work!

FOOD PUZZLES FOR CATS

A food puzzle (also referred to as a puzzle feeder, food toy, or foraging enrichment) is any object that requires interaction for cats to acquire food or treats. They can be easy or complex, depending on your cat's skill level and motivation (or how much of their calendar you need to fill with lunchtime).

Food puzzles come in two general styles: mobile puzzles are balls or other small objects with holes in them that can be rolled or batted around, allowing bits of food to fall out; some of these puzzles even have adjustable holes that can be made larger or smaller to make the puzzle easier or harder for the cat to solve. Stationary puzzles are typically a board with cups, tunnels, or compartments that can hold food. A cat can fish food out of the

different reservoirs with their paws, or open the compartments to access food underneath. You can find these puzzles online or in your favorite pet store (often with the dog toys). Although food puzzles are more convenient to use with dry food, there are plenty of options for wet food or other fresh or homemade diets.

You can also make puzzles for your cats, for example, by cutting holes in a small yogurt container or water bottle and placing some kibbles inside. With a cardboard box, some toilet paper rolls, and a glue gun, you can construct a foraging tray that can keep your cat busy at mealtime. I highly recommend checking out the website foodpuzzlesforcats.com, which I manage with my friend and fellow cat consultant Ingrid Johnson. We've compiled information on just about every type of food puzzle out there, with lots of cute videos of cats using food puzzles, and tips for getting started or making your own puzzles.

CONTRA-WHAT? WILL CATS ACTUALLY WORK FOR FOOD?

Now, some of you might be thinking, "There's no way my cat is going to do any work for their food." The first thing I will say is that you're wrong—cats will work for food. And the second thing I'll say is that my own research suggests that you're not *completely* wrong.

What's more interesting is that several scientific studies have demonstrated that, overall, captive animals actually prefer to work for their food. Over and over again, when given the choice between working for food or eating freely available food that required no additional effort, animals—from pigeons to rats to bears to wolves to humans—consistently choose to exert themselves for treats. This phenomenon is referred to as *contrafreeloading* (the opposite of freeloading).

However, one 1971 study of six laboratory cats found that cats didn't want to work for their food, instead eating all the freely available food before even deigning to lift a paw to get kibbles. Since then, cats have been labeled as the only species that doesn't contrafreeload, giving them a reputation for being lazy and making me wonder if I was misled in recommending that people use food puzzles with their own cats. The study itself was kind of limited in that it was a small sample of cats, and the cats were kept at only 85 percent of their normal body weight (in other words, they were very, very hungry), and they were living in a laboratory. I felt certain that a study of cats living in homes would provide support that cats will actually contrafreeload when the circumstances are ideal.

In 2020, one of my students, Brandon Han, and I ran an experiment to address this obvious gap in our scientific knowledge of cats. Cat owners presented their kitties with the choice of a very basic food puzzle (the Trixie tunnel feeder) or an open tray of the same size and shape as the food puzzle. After acclimating the cats to using the food puzzle, owners conducted ten choice trials with their cat where they were offered the same amount of food on the puzzle and tray, which were offered side by side. The cats could either work for their food by extracting it from the puzzle, or just eat the freely available food from the tray. A video camera was set up so that the cats could eat in the comfort of their own home without experimenter interference, and so we could carefully analyze the data after the fact.

To assess whether cats liked to work for food, we measured what they approached first: the puzzle or the tray? We also recorded which they spent more time at, and which the cat ate more food from. A total of twenty cats (all healthy, spayed or neutered, and living indoors as the only cat in the house) started the study, but three were withdrawn for failure to use the puzzle during the training period.

Although I had high hopes for feline fortitude, the seventeen remaining cats must have read the 1971 paper; they overwhelmingly chose to eat more food from the tray, visited the tray first, and spent more time at the tray. It's not that the puzzle was ignored (although three cats did not eat any food from the puzzle during the study); the cats just clearly took advantage of what was freely available first and really only nibbled from the puzzle. The cats who ate more food from the puzzle also ate more food from the tray, suggesting hunger, rather than a desire to work, was driving the use of the tunnel feeder.

Now, you might be wondering why I'm still telling you to try food puzzles with your cats when the cats *clearly have spoken*! In my own research! The way I like to think about it is: I would typically prefer to sit on the couch and surf the internet, but sometimes I still get myself out of the house to go for a run. The run is good for my short- and long-term health, and although getting started might be tough, at the end I'm usually happy I did it. Although cats may not *prefer* to work for their food, they still may enjoy it and benefit from it. I also have to say that our research (like most research) had some limitations, and it is possible that if we had tried a different type of food puzzle, or used treats instead of the cat's normal diet, the cats may have expressed more interest in contrafreeloading. That's a bunch of future experiments waiting to happen.

But we can use this study to inform how we use food puzzles, being mindful of the fact that (1) some cats just may not take to them and (2) we may have to work to increase a cat's motivation (remember, the food-motivated cats were more likely to use the puzzles). And to be fair, fourteen out of twenty of the cats used the puzzles at least some of the time, so I expect that at least 70 percent of your cats will too.

HOW TO USE FOOD PUZZLES WITH YOUR CAT

Perhaps the most important part of using food puzzles is how you introduce them to your cat. Like they are about most new things, your cat may be cautious or skeptical of this strange device you've placed on the floor. I recommend offering the food puzzle as a choice—either between meals if your cat is meal fed, or as an addendum to their regular food dish if they are living a free-fed lifestyle.

To increase the appeal of the food puzzle, use treats or a novel cat food—either a different brand or different flavor than you usually feed, if your cat is okay with that and there are no health concerns related to switching diets. To give your cat the association of puzzle = food, place some treats near the puzzle, and make sure the puzzle has plenty of food in it! Your cat should barely need to interact with the puzzle at first to get a reward. Once they get the hang of it, your cat will be ready to put a bit more effort into extracting the food (or . . . they just won't!), but at first, we want to make the learning effortless.

Don't give up right away either—you can leave the food puzzle out and available for your cat, and if they choose to use it, that's great. All my cats have enjoyed food puzzles to different degrees, from just for occasional fun with treats to eating all of their dry-food meals from a puzzle (yes, in many cases you can just throw away your food bowls).

Use common sense: If your cat is old and skinny, a food puzzle can be great with treats in addition to their regular meals, but don't make things too challenging if they're not up for it. If your cat is young and food obsessed, they are an excellent candidate for food puzzles. You may even find that they get *too* good at getting the food out of the puzzles, and you may have to get or make more challenging ones to slow them down a little!

INCREASING THE CHALLENGE

If your cat is a food puzzle whiz, there are a few easy ways you can increase the challenge to slow down their feeding and keep them busier longer:

- Use a puzzle with adjustable settings (e.g., with holes that can be made smaller).

- Use or make a puzzle with smaller or fewer openings.
- Place objects (e.g., tissue paper, a Ping-Pong ball) inside the puzzle along with the food so that there are obstacles for the kibbles to move around and to add some weight to the puzzle.
- Use larger rolling puzzles.
- Nest puzzles (i.e., place a small puzzle inside a larger puzzle—for example, a yogurt cup inside of a margarine tub, only placing food in the smaller puzzle).

Only increase the challenge if your cat is ready, willing, and enthusiastic about working for their food.

LIKE THE DINOSAURS, FOOD PUZZLE PLAY CAN BECOME EXTINCT

When your cat interacts with the puzzle, they come to expect a piece of food. In fancy behavior terms, we say that the interaction with the food puzzle is being *reinforced* by the food. Your cat is more likely to interact with the puzzle in the future. What you don't want to happen is for your cat to become frustrated and give up. Make sure the difficulty level of the puzzle matches your cat's ability, and don't let them spend too much time with the puzzle after all the food is gone. If they repeatedly try to interact with the puzzle and don't get the expected food, their interest in the puzzle will wane, in a behavioral phenomenon known as *extinction*. For that reason, it's a good idea to put food puzzles away when they are empty.

IN CASE OF EMERGENCY: FOOD PUZZLE BOYCOTT

If your cat really isn't the food puzzle type, there's no reason to force the issue. In fact, it can be very dangerous if cats go on a hunger strike, even if for just a few days. Rather than let that happen, let's motivate your cat's seeking and hunting behavior for food in other ways. Toss kibbles around for them to chase and eat—this is simple, and honestly, this is the best way to get some cats (especially the tubby ones) moving!

For cats who are fast eaters, you can try a slow-feeder bowl or mat. The bowls usually have partitions, ridges, and patterns that your cat must lick and work around to extract either wet or dry food. The mats are usually made of silicone and, similar to the bowls, have grooves, ridges, and patterns that wet food can be packed into. They require a lot of licking and jaw movements to get all the food out.

I also love snuffle mats, which have a base and several relatively short strips of fabric (usually fleece). They look sort of like a mop-head. After you sprinkle kibbles throughout the mat, your cat must nuzzle and search throughout the fabric for bits of food. Many cats will use both nose and paw to get the food out. The other nice thing about snuffle mats is that you can either buy or make one!

Finally, I also encourage people to split up food into several small portions, hiding dishes in different areas of the house for cats to find. You can use yogurt lids if you need a cheap and easy supply of several "dishes" to place the food on. Hide-and-seek for food is a great way to encourage exploration and activity (in other words, it gets your cat off the couch). This activity also works well when your cat primarily eats a wet-food diet, which can be a bit trickier to use with most (although certainly not all) food puzzles.

SPEAKING OF HIDE-AND-SEEK: NOSEWORK

Nosework was inspired by detection dogs who are trained to search for things like drugs and bombs. It was adapted for pet dogs who could enjoy using their sense of smell without the high stakes associated with finding a dead body or explosives. Nosework training often starts by giving dogs access to a space where small bits of food are hidden in some, but not all, of a collection of small containers. Because the dog can't use visual cues to see if there is food available, they must use their sniffer to find treats.

As dogs become more adept, the real training begins. Nosework trainers usually use specific scents (such as anise and birch) that dogs can discriminate and locate. For example, there might be twenty containers available, but only one contains the scent. There might even be distractions, such as toys or food, to throw the dog's concentration off. When the dog finds the container holding the anise-scented cotton swab, they let their human know (it's called *alerting*). If they are correct, the dog gets a reward.

I knew that nosework was all the rage with my dog-loving friends, but it really didn't occur to me to try it with cats until I met Hanna Fushihara in 2021. She's a dog trainer who had been competing in nosework with her dog. When her dog sadly passed, she needed a new partner, and why not her cat, Muncho? As soon as I saw a video of what Muncho was doing, my mind was blown. Since then, Hanna has taken nosework to the next level with cats!

Hanna first hid small bits of food in a few scattered containers (just like with the dogs) while Muncho was waiting patiently out of sight in his carrier (which he was well trained to love). Once he was released from the carrier, he was free to explore and search the room and containers as he wished. Muncho often stopped to use his excellent sense of smell to assess the moving "scent clouds" and locate multiple treats quickly. He was so focused on the task at hand!

This style of nosework can be a great challenge for many kitties—but by using containers to hold the food bits, you are in a sense limiting your cat's search area to only places where there is a container. Once your cat is used to this exercise, you can skip the container and start hiding bits of food in other places. Using pureed cat treats (such as Inaba Churu) allows you to smear a small amount of food on various surfaces, such as a table leg, the underside of a chair, or the edge of an accessible bookshelf (Hanna recommends placing removable painter's tape on these surfaces first and placing the pureed treat on that—this is cleaner and allows you to remove all traces of scent after each session).

Now your cat can search anywhere in the room for the treats. There are no limitations to the search possibilities, and no visual cues to narrow down the search area. Your cat must rely strictly on olfactory clues.

The nosework task can be made easier or harder by using a different number of containers (or by skipping containers altogether), by using vertical surfaces, or by using more or fewer rewards. If your cat can see you hiding the treats, pretend to place treats in places while they're watching (yes, fake them out!). As your cat gets better at the scent search, encourage them to expand their skills as their nose gets to work.

Scent is also a fascinating modality to use for enrichment because odors move through a space based on several factors, such as humidity and airflow patterns. As molecules move farther away from their source, they become more diffuse and fainter, meaning your cat might have to do more calculations about where those smells originated. Molecules from multiple sources can also combine and create a new olfactory pattern (imagine if you hide a few different types of food—tuna-scent cloud, meet chicken-scent cloud).

Nosework can encourage your cat to explore and to use their nose—two things they would do if they were hunting! But don't take my word for it—you can learn more from Hanna at noseworkcats.com.

YOUR TURN!

A quick way to get started with nosework is to collect a bunch of empty toilet paper rolls. When your cat is out of the room, stand them upright, placing them in any configuration you like. Now place a small treat or kibble in some (but not all) of the toilet paper rolls. Let your cat investigate and explore the rolls, using their sense of smell to find the treats. This game can be made harder or easier by changing the number of toilet paper rolls, the number of treats, and the setup of the rolls!

STAYING AGILE: FELINE AGILITY

Agility is a sport that was originally modeled after horse jumping, and like many fun activities, dogs seemed to be invited to participate first. Dog agility got its start in the 1970s, and since then has only grown in popularity. Over a million dogs participate in the American Kennel Club's agility competitions each year. There are multiple international competitions, and there are even world competitions for dog handlers under the age of eighteen. Regardless of location, agility courses are created from a series of obstacles, such as hurdles, seesaws, tunnels, hoops, and A-frames to climb over. The layout of the obstacles can vary at each event based on the judges' whims. It's kind of like *American Ninja Warrior*, but for dogs. Handlers must lead their dogs through the course without touching the dogs or any of the obstacles, and no food or toy incentives are allowed. The dog who completes the course with the best time and accuracy wins.

A little late to the party, feline agility took off around twenty years ago. Just like the dogs, cats are trained (believe it or not) to race through an obstacle course of items such as hoops, ramps, and tunnels. Interest in cat agility has exploded to the point that now there are even agility tournaments exclusively for cats.

The standard agility course features steps, hurdles ranging in height from one to four bars, two hoops, two tunnels, and a series of weave poles. Cats are generally trained by luring them through the various obstacles. During competition, which usually happens at a regional cat show, wand toys are allowed, but treats and food are not. And no touching the cat, please! The cat must complete the agility course with an audience, and so agility competitions are best suited to confident cats who aren't put off by being in a new place, surrounded by strangers (and other cats waiting in the wings for their turn to compete).

The cats are timed, from the moment they touch the first obstacle until they make it through the very last obstacle. Cats get points for each obstacle they complete, with bonus points for a "clean run"—meaning that they complete the entire course in order, in a clockwise fashion, and without any feline distractions, within four and a half minutes. Although not all cats make it through (or, for some cats, even get started), a good time is had by all (especially the spectators, but maybe not the disappointed owner who was hoping for a ribbon).

If you think your cat has a shot at the big leagues, the Cat

Fanciers' Association has a thorough guide to cat agility on their website, and you can also check out the International Cat Agility Tournaments website at catagility.com. However, to do agility with your own cat at home, you don't have to be this formal in your approach. You can buy or build obstacles at home, creating your own course for your cat to run on or around. A chair, a Hula-Hoop, a crinkle tunnel, and a scratching post can easily become a mini agility course! If your cat is going to be jumping over or through anything (like that Hula-Hoop), be sure to start low to the ground and gradually increase the difficulty as your cat gets more comfortable. If these objects are too high at first, your cat is more likely to slip under them rather than jump over them!

Although formal agility competitions do not allow food, there's no reason not to use treats to get your cat started with your agility course at home. Toys can also be an excellent lure to get your cat to go over a jump or through a tunnel. There are plenty of amazing videos online to get you inspired, and the most important thing to remember is that it should be fun for you . . . and your cat.

THE CHALLENGES OF GOING SOLO

One thing I ask most of my clients is how often they play with their cat. Their response inevitably is "Oh, my cat has toys everywhere," as if that is somehow an indicator of *their* interactions with their cat. This sentence is almost always followed by the phrase "which he completely ignores."

There's nothing wrong with solo toys per se, except that they are dead and lifeless on their own. They require our cats to use a lot of imagination and motivation to make them move and behave like prey. For kittens, this is quite easy—to them, anything that slightly resembles prey is worth investigation. For adult cats, it's a bit harder for them to use their imaginations; the motivation to test things and experiment with them (especially things that are dusty and have been sitting on the floor in the middle of the living room for two weeks) is much lower. If solo toys are your cat's main source of entertainment, they might not find life particularly entertaining.

I'm an only child. For better or worse, I spent a lot of time by myself as a child. This included playing board games like

Monopoly or Life by myself. I would play for me and my competitor (but hey, I always won!). I'd roll the dice, move the pieces, and strategize, trying to pretend I didn't know exactly what I was going to do when I picked up the other pawn.

I liken solo toys for your cat to those board games I played by myself. When you provide your cat with those small toys, they must both infuse the life into them and then pounce on them. But they already know the force and trajectory of how those toys will move and how far away they will land. They know how the toy will feel under their paws and between their teeth. And in their heart of hearts, they know that toy poses no threat: the wind has blown out of those predatory sails.

The whole point of this book has been allowing your cat to lose themselves in the hunting experience; we do that by playing the part of the prey, by animating the toys. It's not that we shouldn't provide solo toys for our cats; I truly believe they can be a great source of fun and enrichment. We just need to use them thoughtfully and as part of a larger enrichment plan, not as the main course of a meal.

TIPS FOR SOLO TOY USE

1. Safety first! Inspect toys for any choking hazards or bits that could be chewed off.

2. Have a variety of toys, providing some that are small, lightweight, mobile (e.g., can roll), and have

different textures (e.g., fur mice, Ping-Pong balls, spongy soccer balls, sparkle puffs, crinkle balls). Think prey-like items that can be easily manipulated.

3. Rotate toys frequently so they don't become stale and boring. Once or twice a week pick up all the small toys and place them in a storage bin. Select some toys that differ in style, size, or texture to place out for the next several days.

4. Don't be shy about throwing those "solo" toys around for your cat to chase. Some cats enjoy fetch, and there's nothing wrong with tossing a few toys in the air to get your cat moving.

5. Vary your movements—try rolling or skipping (like a stone) some toys, try a good ol' overhand or underhand toss with the others, or try throwing them straight in the air. See what gets your cat excited.

6. Know your cat—choose toys and movements that will be interesting but not scary or overwhelming to them.

FROM LOW-TECH TO HIGH-TECH

Now it's time to delve into what I jokingly call the "neglect-o-matic" toys (I think I first heard that term after my friends had a baby and purchased an automatic baby-soothing device). These are toys that might have some level of sustained movement, whether instigated by the cat or by a switch that you flip on.

The most basic type of self-play toy features a ball in some sort of track. The cat pushes the ball, and the ball rolls around in a set pathway. Perhaps it's a circular track or a multilevel track that a ball slowly rolls down. I've even seen some that resemble the Hot Wheels track of my childhood dreams. But let's not kid ourselves—this is a toy with limited appeal. My experience is that kittens are most likely to use this toy, although some adult cats may give the ball a little push every now and again. When I had a round track toy with a scratcher in the middle, it would sit unused for months. But I swear, for years, every time I had a notion to put that track toy out in the trash bin, one of my cats—like they could read my mind—would suddenly show a marked curiosity about it. So it lingered in our living room (I did eventually get rid of it, although I sometimes question this decision).

We can take things up a notch with robotic toys. People are very drawn to these toys because they seem like the perfect answer to all of our playtime problems. The robot does the vacuuming, a machine washes our dishes, can't we use technology to our advantage and take care of our cat's needs?

The answer is complicated. Certainly a lot of tech companies have been founded on that very idea—there is a problem to be solved—my cat needs play but I don't wanna! The toy companies promise a solution—a giant mouse that rolls around, a fish that flops endlessly, a base with a stick and a feather that whips around in a frenzy. The possibilities have proven endless as robot toy after robot toy has hit the market.

Cats do not have a reputation for being picky for no reason. My advice to anyone considering robotic toys is: Buyer beware! One of the biggest issues I have with robotic toys is that for the toy to move, there must be a motor of some sort. The toy may also need to house batteries. To fit these things inside, that usually means the toy is somewhat large. With size and a motor comes noise. Noise that may be intimidating to your cat. The toys also do not tend to have realistic movement. It's as if translating movement to ones and zeros has led to something being lost—that something being a similarity to the ways actual prey animals move. Loud, scary, awkward movements are not going to get any cat excited to play.

That said, some cats do enjoy these toys, and a few of them are actually pretty good—especially when used in moderation. They should never be the mainstay of your play routine with your cat, but they can definitely supplement it. My advice: Watch videos of the toy in motion and see if it sounds and looks like something your cat would be comfortable with. Is it loud when rolling against a hardwood floor? Is it too fast? How do cats in the video respond? Are they showing the classic hunting moves, like the butt wiggle and pounce, or are they swatting at the toy as if to say, "Stay back"? And be prepared to spend (waste) some money on some of these automatic toys, which may or may not be a hit with your own cats.

My personal favorites are those that are small and quiet, such as the Hexbug nano and similar knockoffs—just a tiny bug with a feathery tail that has a quiet buzz and semi-random movements. My cats also enjoy playing with water, and the small fish that are made for aquatic play have been a hit in our household. They light up and "swim" around when they are placed in water. I also recommend trying toys that involve anything moving under fabric, or a toy that gently pops in and out of hiding—preferably toys that have a "slow" setting, which is more likely to be enticing and nonthreatening to your cat.

On a whim, I got a bird drone a few years back. The body of the bird is made of Styrofoam, and it has plastic wings that flutter and flap. The controller is more of a "barely-in-control-er," and the drone acts more like a confused bird that accidentally got trapped in your house than a real bird, although it does elicit curiosity (and maybe some confusion) in my feline crew. Use with caution.

GUIDE FOR USING ROBOTIC TOYS

- Purchase thoughtfully, keeping your cat's play preferences in mind.
- Test the toy out with your cat by placing it on the floor several feet away, allowing your cat to interact with the robotic toy if they feel comfortable.
- If your cat seems afraid, that toy may not work out for your home.

- Automatic toys can only supplement interactive play and should not be the main form of enrichment for your cat.
- Like with other types of toys, automatic toys should be rotated in and out of your enrichment schedule so they do not become boring.

FOR THE COUCH POTATOES

When is your cat old enough for a cell phone? How much gaming is too much for a cat? Luckily, we're only asking these questions about children right now, but it is possible that in the not-so-distant future, we'll be considering whether our cats are becoming "digital junkies." Since their ability to surf the internet is pretty limited, I'm not too worried right now. That said, a lot of people are curious about whether video games made for cats are a good idea.

Video games for cats are typically simple programs that feature animated "prey" that can move around the screen. If you're using a touch-sensitive screen, your cat may even be able to earn points when they bat at the moving objects. Whether your cat will be interested in playing video games depends on a few things: the quality of the game, your cat's interest in this type of visual stimulation, and how frustrated your cat gets with a two-dimensional game that doesn't allow for much payoff.

For some cats, playing on a tablet or phone is just too close-up to really see what is going on. They can discriminate some movements, but a fancy-looking game will likely not be any more enticing than a very simple one (like *Pong*!). In addition, they won't receive any of the types of feedback they would get from a close-up prey item or toy (such as sensing air movements or touch with their paws and whiskers). The fun and interest may be short-lived for most cats. If your cat seems to enjoy video games, keep in mind that they will be pawing your tablet or phone screen and may even knock it off whatever it is resting on. I highly recommend supplementing any video play with some time with a real toy (much like if you are going to use a laser pointer) to prevent frustration.

Cat TV is a more passive form of watching, as no pawing is required and no points may be acquired. Some cats seem to naturally gravitate toward movement on a television screen (such as a tennis match), and others seem to barely register it. The cats who do respond to television may be more sensitive to two-dimensional movements, and may be even more interested when bird or mouse sounds are added (such as in many of the videos designed specifically for cat entertainment). A study of twenty-five cats in an animal shelter found that when videos were played for them, they indulged for only about 6 percent of their day, but they did spend more time looking at TV screens when they featured animation or movement compared to just a blank screen. Although your cat may enjoy a small amount of television time, they aren't as likely to find those mouse videos bingeworthy; as I previously mentioned, videos made for cats might be better used to get your cat in the mood for play rather than relied upon for providing any sort of hunting experience in and of themselves.

♥ ♥ ♥

Whether it's encouraging your cat to work for their food, use their nose, jump through hoops, or just kick back on the couch with an iPad, we have lots of options for expanding their predatory horizons. Like everything else about play, it may take some trial and error to find what your cat responds to best. Luckily, like everything else about play, the journey itself is just as fun as the destination!

Epilogue

I wanted to shine a spotlight on play to help cats and humans have a better connection. Playtime leads to a healthier cat who is less likely to have behavior problems. Playtime increases a human's investment in their cat's happiness. Playtime leads to a stronger human-cat relationship, through the shared activity and the routine and positive experiences that playtime creates.

Although I have written what is in many respects a prescriptive "how-to," this is not an operator's manual. Instead, I hope I left you with a feeling of awe at your cat's prowess and predatory talents; with excitement about how you can tap into that predatory instinct through play; with the skills to execute and improvise a play session; and with a sense of joy knowing that even if you haven't figured out every last thing about your cat, by knowing this *one thing*—how to play with them—you can have a more complete relationship with your cat.

Acknowledgments

This book would not exist without two people: Julie Hecht and Lili Chin. It was through collaborations with these amazing and talented women that I realized there was a book about cat play that needed to be written. I feel so lucky to call you both dear friends.

I want to thank Scott Bradley for being the kindest and most supportive partner who also fully participates in our daily play sessions with the cats. Mary Jane Weatherbee not only read the entire first draft of the book but also provided me with endless moral support—you are a dear friend and I owe you.

I would not be where I am today if not for meeting Dilara Göksel Parry many years ago. You opened the world of cat behavior to me and I am eternally grateful! I want to give a thank-you and big hug to Jackson Galaxy for your friendship, for dragging me along on your book writing journey, and for supporting mine. Hannah Shaw makes the world a better place for humans and other animals; thank you for always looking out for your friends, including me. And I am grateful to my mother, Roberta, for always cheerleading my nontraditional adulthood.

Gillian MacKenzie was incredibly patient with me while I figured out what my book was going to be about. Thank you for taking a chance on me. A special shout-out to my editors, Lauren Appleton and Ashley Alliano, for helping me express my ideas clearly. You both made the editing process relatively painless, and the result is a better book. Thank you to Kate Johnson at Gillian MacKenzie Agency, and everyone at TarcherPerigee and Profile Books. I would also like to thank Sara Carder for believing in this book!

Hanna Fushihara and Dr. Greg Gbur both generously gave their time to chat with me about their respective areas of expertise, which I greatly appreciate! There are many other people who offered guidance, support, or friendship during the time I wrote this book, including but not limited to Hal Herzog, Eleanor Spicer Rice, John Bradshaw, Tony Buffington, Judi Stella, Ingrid Johnson, and Tracie Hotchner. Thank you all.

Finally, without cats, there would be no playing with cats. So thank you, cats, for existing and fueling my lifelong obsession.

Bibliography

Albonetti, M. E. 1988. "Behavioural Development in Kittens: Effects of Litter Sex-Ratio." *Monitore Zoologico Italiano-Italian Journal of Zoology* 22: 53–61.

Barrett, P., and P. Bateson. 1978. "The Development of Play in Cats." *Behaviour* 66: 106–20.

Bateson, P., and M. Young. 1979. "The Influence of Male Kittens on the Object Play of Their Female Siblings." *Behavioral and Neural Biology* 27: 374–78.

Bateson, P., and M. Young. 1981. "Separation from the Mother and the Development of Play in Cats." *Animal Behaviour* 29: 173–80.

Bekoff, M., and J. A. Byers, eds. 1998. *Animal Play: Evolutionary, Comparative and Ecological Perspectives.* Cambridge, UK: Cambridge University Press.

Belsito, K. R. et al. 2009. "Impact of Ovariohysterectomy and Food Intake on Body Composition, Physical Activity, and Adipose Gene Expression in Cats." *Journal of Animal Science* 87: 594–602.

Bennett, P. C., N. J. Rutter, J. K. Woodhead, and T. J. Howell. 2017. "Assessment of Domestic Cat Personality, as Perceived by 416 Owners, Suggests Six Dimensions." *Behavioural Processes* 141: 273–83.

Biben, M. 1979. "Predation and Predatory Play Behaviour of Domestic Cats." *Animal Behaviour* 27: 81–94.

Biello, D. 2007. "Strange but True: Cats Cannot Taste Sweets." *Scientific American.*

Birch, A. M., and Á. M. Kelly. 2019. "Lifelong Environmental Enrichment in the Absence of Exercise Protects the Brain from Age-Related Cognitive Decline." *Neuropharmacology* 145: 59–74.

Blois-Heulin, C. et al. 2015. "Animal Welfare: Could Adult Play Be a False Friend?" *Animal Behavior and Cognition* 2: 156–85.

Bol, S. et al. 2017. "Responsiveness of Cats (*Felidae*) to Silver Vine (*Actinidia polygama*), Tatarian Honeysuckle (*Lonicera tatarica*), Valerian (*Valeriana officinalis*) and Catnip (*Nepeta cataria*)." *BMC Veterinary Research* 13: 1–16.

Bradshaw, J. W. 2012. *The Behaviour of the Domestic Cat.* Centre for Agriculture and Bioscience International: Wallingford, UK.

Bradshaw, J. W. 2016. "Sociality in Cats: A Comparative Review." *Journal of Veterinary Behavior* 11: 113–24.

Bradshaw, J. W., and S. L. Hall. 1999. "Affiliative Behaviour of Related and Unrelated Pairs of Cats in Catteries: A Preliminary Report." *Applied Animal Behaviour Science* 63: 251–55.

Bradshaw J. W. S., and R. E. Lovett. 2003. "Dominance Hierarchies in Domestic Cats: Useful Construct or Bad Habit?" *Proceedings of the British Society of Animal Science Conference*: 16.

Burghardt, G. M. 2011. "Defining and Recognizing Play." In *The Oxford Handbook of the Development of Play*. Edited by A. D. Pellegrini. New York: Oxford University Press, 9–18.

Calver, M., S. Thomas, S. Bradley, and H. McCutcheon. 2007. "Reducing the Rate of Predation on Wildlife by Pet Cats: The Efficacy and Practicability of Collar-Mounted Pounce Protectors." *Biological Conservation* 137: 341–48.

Caro, T. M. 1980. "Effects of the Mother, Object Play, and Adult Experience on Predation in Cats." *Behavioral and Neural Biology* 29: 29–51.

Caro, T. M. 1980. "The Effects of Experience on the Predatory Patterns of Cats." *Behavioral and Neural Biology* 29: 1–28.

Caro, T. M. 1981. "Sex Differences in the Termination of Social Play in Cats." *Animal Behaviour* 29: 271–79.

Cecchetti, M., S. L. Crowley, C. E. D. Goodwin, and R. A. McDonald. 2021. "Provision of High Meat Content Food and Object Play Reduce Predation of Wild Animals by Domestic Cats *Felis catus*." *Current Biology* 31: 1107–11.

Clarke, D. L. et al. 2005. "Using Environmental and Feeding Enrichment to Facilitate Feline Weight Loss." *Journal of Animal Physiology and Animal Nutrition* 89: 427–33.

Crowell-Davis, S. L., T. M. Curtis, and R. J. Knowles. 2004. "Social Organization in the Cat: A Modern Understanding." *Journal of Feline Medicine and Surgery* 6: 19–28.

Dantas, L. M. S., M. M. Delgado, I. Johnson, and C. T. Buffington. 2016. "Food Puzzles for Cats: Feeding for Physical and Emotional Wellbeing." *Journal of Feline Medicine and Surgery* 18: 723–32.

de Godoy, M. R. C. et al. 2015. "Feeding Frequency, but Not Dietary Water Content, Affects Voluntary Physical Activity in Young Lean Adult Female Cats." *Journal of Animal Science* 93: 2597–601.

de Godoy, M. R., and A. K. Shoveller. 2017. "Overweight Adult Cats Have Significantly Lower Voluntary Physical Activity Than Adult Lean Cats." *Journal of Feline Medicine and Surgery* 19: 1267–73.

Delgado, M., and L. M. S. Dantas. 2020. "Feeding Cats for Optimal Mental and Behavioral Well-Being." *Veterinary Clinics of North America: Small Animal Practice* 50: 939–53.

Delgado, M. M., B. S. G. Han, and M. J. Bain. 2021. "Domestic Cats (*Felis catus*) Prefer Freely Available Food over Food That Requires Effort." *Animal Cognition* 25: 95–102.

Deng, P. et al. 2014. "Effects of Feeding Frequency and Dietary Water Content on Voluntary Physical Activity in Healthy Adult Cats." *Journal of Animal Science* 92: 1271–77.

Dickman, C. R., and T. M. Newsome. 2015. "Individual Hunting Behaviour and Prey Specialisation in the House Cat *Felis Catus*: Implications for

Conservation and Management." *Applied Animal Behaviour Science* 173: 76–87.

Driscoll, C. A., D. W. Macdonald, and S. J. O'Brien. 2009. "From Wild Animals to Domestic Pets, an Evolutionary View of Domestication." *Proceedings of the National Academy of Sciences* 106: 9971–78.

Duffy, D. L., R. T. D. de Moura, and J. A. Serpell. 2017. "Development and Evaluation of the Fe-BARQ: A New Survey Instrument for Measuring Behavior in Domestic Cats (*Felis s. catus*)." *Behavioural Processes* 141: 329–41.

Ellis, J. J., H. Stryhn, J. Spears, and M. S. Cockram. 2017. "Environmental Enrichment Choices of Shelter Cats." *Behavioural Processes* 141: 291–96.

Ellis, S. L. H., and D. L. Wells. 2008. "The Influence of Visual Stimulation on the Behaviour of Cats Housed in a Rescue Shelter." *Applied Animal Behaviour Science* 113: 166–74.

Ellis, S. L. H. et al. 2013. "AAFP and ISFM Feline Environmental Needs Guidelines." *Journal of Feline Medicine and Surgery* 15: 219–30.

Elzerman, A. L., T. L. DePorter, A. Beck, and J.-F. Collin. 2020. "Conflict and Affiliative Behavior Frequency Between Cats in Multi-Cat Households: A Survey-Based Study." *Journal of Feline Medicine and Surgery* 22: 705–17.

Evans, R., M. Lyons, G. Brewer, and E. Bethell. 2021. "A Domestic Cat (*Felis silvestris catus*) Model of Triarchic Psychopathy Factors: Development and Initial Validation of the CAT-Tri+ Questionnaire." *Journal of Research in Personality* 95: 104161.

Fraser, A. F. 2012. *Feline Behaviour and Welfare*. Centre for Agriculture and Bioscience International: Wallingford, UK.

Gajdoš-Kmecová, N. et al. 2023. "An Ethological Analysis of Close-Contact Inter-Cat Interactions Determining If Cats Are Playing, Fighting, or Something in Between." *Scientific Reports* 13: 1–11.

Galvan, M., and J. Vonk. 2016. "Man's Other Best Friend: Domestic Cats (*F. silvestris catus*) and Their Discrimination of Human Emotion Cues." *Animal Cognition* 19: 193–205.

Golle, J., S. Lisibach, F. W. Mast, and J. S. Lobmaier. 2013. "Sweet Puppies and Cute Babies: Perceptual Adaptation to Babyfacedness Transfers Across Species." *PloS One* 8: e58248.

Gosling, S. D., and O. P. John. 1999. "Personality Dimensions in Nonhuman Animals: A Cross-Species Review." *Current Directions in Psychological Science* 8: 69–75.

Hall, S. L. 1998. "Object Play by Adult Animals." In *Animal Play: Evolutionary, Comparative, and Ecological Perspectives*. Edited by M. Bekoff and J. A. Byers. Cambridge, UK: Cambridge University Press, 45–60.

Hall, S. L., and J. W. S. Bradshaw. 1998. "The Influence of Hunger on Object Play by Adult Domestic Cats." *Applied Animal Behaviour Science* 58: 143–50.

Hall, S. L., J. W. S. Bradshaw, and I. H. Robinson. 2002. "Object Play in Adult Domestic Cats: The Roles of Habituation and Disinhibition." *Applied Animal Behaviour Science* 79: 263–71.

Held, S. D. E., and M. Špinka. 2011. "Animal Play and Animal Welfare." *Animal Behaviour* 81: 891–99.

Hernandez, S. M. et al. 2018. "Activity Patterns and Interspecific Interactions of Free-Roaming, Domestic Cats in Managed Trap-Neuter-Return Colonies." *Applied Animal Behaviour Science* 202: 63–68.

Herrera, D. J. et al. 2022. "Prey Selection and Predation Behavior of Free-Roaming Domestic Cats (*Felis catus*) in an Urban Ecosystem: Implications for Urban Cat Management." *Biological Conservation* 268: 109503.

Houser, B., and K. R. Vitale. 2022. "Increasing Shelter Cat Welfare Through Enrichment: A Review." *Applied Animal Behaviour Science* 248: 105585.

Jacobs, B. L., L. O. Wilkinson, and C. A. Fornal. 1990. "The Role of Brain Serotonin: A Neurophysiologic Perspective." *Neuropsychopharmacology* 3: 473–79.

Kaplan, G. 2020. "Play Behaviour, Not Tool Using, Relates to Brain Mass in a Sample of Birds." *Scientific Reports* 10: 1–15.

Katz, R. J. 1980. "Role of Serotonergic Mechanisms in Animal Models of Predation." *Progress in Neuro-Psychopharmacology* 4: 219–31.

Koffer, K., and G. Coulson. 1971. "Feline Indolence: Cats Prefer Free to Response-Produced Food." *Psychonomic Science* 24: 41–42.

Larsen, J. A. 2017. "Risk of Obesity in the Neutered Cat." *Journal of Feline Medicine and Surgery* 19: 779–83.

Lascelles, B. D. X. et al. 2008. "Evaluation of a Digitally Integrated Accelerometer-Based Activity Monitor for the Measurement of Activity in Cats." *Veterinary Anaesthesia and Analgesia* 35: 173–83.

Levine, E., P. Perry, J. Scarlett, and K. A. Houpt. 2005. "Intercat Aggression in Households Following the Introduction of a New Cat." *Applied Animal Behaviour Science* 90: 325–36.

Leyhausen, P. 1979. *Cat Behavior: The Predatory and Social Behavior of Domestic and Wild Cats.* New York: Garland STPM Press.

Li, X. et al. 2006. "Cats Lack a Sweet Taste Receptor." *Journal of Nutrition* 136: 1932S–934S.

Loberg, J. M., and F. Lundmark. 2016. "The Effect of Space on Behaviour in Large Groups of Domestic Cats Kept Indoors." *Applied Animal Behaviour Science* 182: 23–29.

Lowe, S. E., and J. W. S. Bradshaw. 2001. "Ontogeny of Individuality in the Domestic Cat in the Home Environment." *Animal Behaviour* 61: 231–37.

Loyd, K. A. T. et al. 2013. "Quantifying Free-Roaming Domestic Cat Predation Using Animal-Borne Video Cameras." *Biological Conservation* 160: 183–89.

Martell-Moran, N. K., M. Solano, and H. G. Townsend. 2018. "Pain and Adverse Behavior in Declawed Cats." *Journal of Feline Medicine and Surgery* 20: 280–88.

Martin, P. 1984. "The Time and Energy Costs of Play Behaviour in the Cat." *Zeitschrift für Tierpsychologie* 64: 298–312.

Martin, P., and P. Bateson. 1985. "The Ontogeny of Locomotor Play Behaviour in the Domestic Cat." *Animal Behaviour* 33: 502–10.

McCune, S. 1995. "The Impact of Paternity and Early Socialisation on the Development of Cats' Behaviour to People and Novel Objects." *Applied Animal Behaviour Science* 45: 109–24.

McGregor, H., S. Legge, M. E. Jones, and C. N. Johnson. 2015. "Feral Cats Are Better Killers in Open Habitats, Revealed by Animal-Borne Video." *PLoS One* 10: e0133915.

McMillan, F. D. 2013. "Stress-Induced and Emotional Eating in Animals: A Review of the Experimental Evidence and Implications for Companion Animal Obesity." *Journal of Veterinary Behavior* 8: 376–85.

Mendl, M. 1988. "The Effects of Litter-Size Variation on the Development of Play Behaviour in the Domestic Cat: Litters of One and Two." *Animal Behaviour* 36: 20–34.

Mendoza, D. L., and J. M. Ramirez. 1987. "Play in Kittens (*Felis domesticus*) and Its Association with Cohesion and Aggression." *Bulletin of the Psychonomic Society* 25: 27–30.

Mertens, C., and D. C. Turner. 1988. "Experimental Analysis of Human-Cat Interactions During First Encounters." *Anthrozoös* 2: 83–97.

Michel, K., and M. Scherk. 2012. "From Problem to Success: Feline Weight Loss Programs That Work." *Journal of Feline Medicine and Surgery* 14: 327–36.

Moore, A. M., and M. J. Bain. 2013. "Evaluation of the Addition of In-Cage Hiding Structures and Toys and Timing of Administration of Behavioral Assessments with Newly Relinquished Shelter Cats." *Journal of Veterinary Behavior* 8: 450–57.

Pellis, S. M. et al. 1988. "Escalation of Feline Predation Along a Gradient from Avoidance Through 'Play' to Killing." *Behavioral Neuroscience* 102: 760–77.

Piccione, G. et al. 2013. "Daily Rhythm of Total Activity Pattern in Domestic Cats (*Felis silvestris catus*) Maintained in Two Different Housing Conditions." *Journal of Veterinary Behavior* 8: 189–94.

Pyari, M. S., S. Uccheddu, R. Lenkei, and P. Pongrácz. 2021. "Inexperienced but Still Interested—Indoor-Only Cats Are More Inclined for Predatory Play Than Cats with Outdoor Access." *Applied Animal Behaviour Science* 241: 105373.

Rowe, E. et al. 2015. "Risk Factors Identified for Owner-Reported Feline Obesity at Around One Year of Age: Dry Diet and Indoor Lifestyle." *Preventive Veterinary Medicine* 121: 273–81.

Rowe, E. C. et al. 2017. "Early-Life Risk Factors Identified for Owner-Reported Feline Overweight and Obesity at Around Two Years of Age." *Preventive Veterinary Medicine* 143: 39–48.

Salonen, M. et al. 2019. "Breed Differences of Heritable Behaviour Traits in Cats." *Scientific Reports* 9: 1–10.

Shmalberg, J. 2019. "Role of Exercise in the Management of Obesity." In *Obesity in the Dog and Cat*. Edited by M. G. Cline and M. Murphy. Boca Raton, FL: CRC Press, 121–31.

Shreve, K. R. V., L. R. Mehrkam, and M. A. R. Udell. 2017. "Social Interaction, Food, Scent or Toys? A Formal Assessment of Domestic Pet and Shelter Cat (*Felis silvestris catus*) Preferences." *Behavioural Processes* 141: 322–28.

Shyan-Norwalt, M. R. 2005. "Caregiver Perceptions of What Indoor Cats Do 'For Fun.'" *Journal of Applied Animal Welfare Science* 8: 199–209.

Sinn, L. 2016. "Factors Affecting the Selection of Cats by Adopters." *Journal of Veterinary Behavior* 14: 5–9.

Stella, J. L., L. K. Lord, and C. T. Buffington. 2011. "Sickness Behaviors in Response to Unusual External Events in Healthy Cats and Cats with Feline Interstitial Cystitis." *Journal of the American Veterinary Medical Association* 238: 67–73.

Stella, J., C. Croney, and T. Buffington. 2013. "Effects of Stressors on the Behavior and Physiology of Domestic Cats." *Applied Animal Behaviour Science* 143: 157–63.

Strickler, B. L., and E. A. Shull. 2014. "An Owner Survey of Toys, Activities, and Behavior Problems in Indoor Cats." *Journal of Veterinary Behavior* 9: 207–14.

Tarou, L. R., and M. J. Bashaw. 2007. "Maximizing the Effectiveness of Environmental Enrichment: Suggestions from the Experimental Analysis of Behavior." *Applied Animal Behaviour Science* 102: 189–204.

Thomson, J. E., S. S. Hall, and D. S. Mills. 2018. "Evaluation of the Relationship Between Cats and Dogs Living in the Same Home." *Journal of Veterinary Behavior* 27: 35–40.

Vinke, C. M., L. M. Godijn, and W. J. R. van der Leij. 2014. "Will a Hiding Box Provide Stress Reduction for Shelter Cats?" *Applied Animal Behaviour Science* 160: 86–93.

West, M. 1974. "Social Play in the Domestic Cat." *American Zoologist* 14: 427–36.

West, M. J. 1977. "Exploration and Play with Objects in Domestic Kittens." *Developmental Psychobiology* 10: 53–57.

Willson, S. K., I. A. Okunlola, and J. A. Novak. 2015. "Birds Be Safe: Can a Novel Cat Collar Reduce Avian Mortality by Domestic Cats (*Felis catus*)?" *Global Ecology and Conservation* 3: 359–66.

About the Author

Mikel Maria Delgado, PhD, is looking to make the world a better place for cats. She is a cat behavior consultant at Feline Minds, where she offers assistance to cat owners, animal shelters, and corporations. Delgado completed her PhD in animal behavior and cognition at UC Berkeley and was a postdoctoral fellow at the UC Davis School of Veterinary Medicine, where she researched the social and feeding behaviors of cats and the health and behavior of orphaned neonatal kittens. She coauthored the 2017 book *Total Cat Mojo* with Jackson Galaxy. Delgado lives in Sacramento, California, with her boyfriend, Scott, and their three rescue cats: Coriander, Ruby, and Professor Scribbles.

Illustrator **Lili Chin** has spent the past fifteen years drawing dogs and cats for animal lovers and educators. She is the author of two books, *Doggie Language* and *Kitty Language.*